Forests for People

Forests for People

Community Rights and Forest Tenure Reform

Edited by

Anne M. Larson, Deborah Barry,
Ganga Ram Dahal
and Carol J. Pierce Colfer

publishing for a sustainable future
London • Washington, DC

First published in 2010 by Earthscan

Earthscan Ltd, Dunstan House, 14a St Cross Street, London EC1N 8XA, UK
Earthscan LLC, 1616 P Street, NW, Washington, DC 20036, USA
Earthscan publishes in association with the International Institute for Environment and Development

For more information on Earthscan publications, see www.earthscan.co.uk or write to earthinfo@earthscan.co.uk

ISBN: 978-1-84407-917-9 hardback
ISBN: 978-1-84407-918-6 paperback

Typeset by JS Typesetting Ltd, Porthcawl, Mid Glamorgan
Cover design by Susanne Harris

A catalogue record for this book is available from the British Library

Library of Congress Cataloging-in-Publication Data

Forests for people : community rights and forest tenure reform / edited by Anne M. Larson ... [et al.].
p. cm.
Includes bibliographical references and index.
ISBN 978-1-84407-917-9 (hardback) – ISBN 978-1-84407-918-6 (pbk.)
1. Community forestry–Developing countries. 2. Forest policy–Developing countries. 3. Land tenure–Developing countries. 4. Forest management–Citizen participation–Economic aspects–Developing countries. I. Larson, Anne M.
SD669.5.F67 2010
333.7509172'4–dc22
2009033828

At Earthscan we strive to minimize our environmental impacts and carbon footprint through reducing waste, recycling and offsetting our CO_2 emissions, including those created through publication of this book. For more details of our environmental policy, see www.earthscan.co.uk.

Printed and bound in the UK by The Cromwell Press Group.
The paper used is FSC certified.

Contents

List of Figures, Tables and Boxes

Figures

Tables

Boxes

Foreword

There are many reasons besides conservation of forests and biodiversity to be concerned with the status of the world's forests, ranging from poverty alleviation to cultural preservation of indigenous communities to global climate change. Yet up until now there has been little attempt to synthesize what is known about efforts to grant new tenure rights to communities living in or near forests, or what has come to be known as 'forest reform'. Since 2002 alone 15 of the 30 most forested countries worldwide have increased the forest area available for use, management or ownership by local communities. The authors of *Forests for People* argue that a number of factors have combined so that the rights of these communities are finally being recognized. Among the reasons for this global trend is the growing recognition that conservation, sustainability and enhanced livelihoods for those who have traditionally depended upon the forests may be complementary goals.

This book represents the culmination of a three-year study of ten countries in three regions of the world – Africa, Asia and Latin America – funded by International Development Research Centre (IDRC) and the Ford Foundation, with additional support in Latin America from Program on Forests (PROFOR)/World Bank. The international team, from the Center for International Forestry Research (CIFOR), joined forces with national and local researchers to investigate the process of reform in over 30 research sites, chosen because a change in tenure status had already occurred or was about to occur. The sites were also chosen because of the opportunity they represented to deepen the reform process in favour of local communities or to dialogue with policy-makers over the efficacy of the reforms. The authors make no bones about their own agenda: to promote the local control and management of forests, or community forest management.

The main questions addressed include the impetus for and nature of the reforms, the key mediating factors influencing outcomes – specifically, the role of local organizations and regulators and regulations, as well as markets – and the specific outcomes of the reforms for livelihoods, condition of the forests and equity. The central argument of the book is that community forest management can serve complementary aims. It also provides a rather high standard against which to measure the success of the reforms.

This book does a superb job of synthesizing the lessons learned. Rather than presenting a series of country case studies as is often the case in comparative volumes spanning three continents, the chapters are organized according to cross-cutting themes. Each chapter draws upon the findings for a range of countries, sometimes in all three regions, and includes a literature review.

Forest reform differs significantly from the agrarian reforms of the past in that the rights ceded are collective, rather than individual, and exclude the right to alienation. Rather than involving a redistribution of land, the primary beneficiaries are usually those who already live in the forests or whose livelihoods depend upon them. Moreover, the state tends to maintain a relatively large role in management to ensure that forests remain intact. The range of new rights transferred to communities varies tremendously, with the most effective reforms being those that grant not only use, but also exclusion rights.

This book makes a very convincing case regarding the potential benefits of community forest management. It also makes clear the many obstacles faced by real tenure reform efforts. Each phase of the reform process has faced different challenges. The most ambitious reforms are associated with strong and effective pressure from the grassroots, such as the demands for demarcation of indigenous territories in Latin America. The implementation phase has faced obstacles everywhere, largely associated with the regulatory framework of the reform processes as well as with the weaknesses of the community governance structures needed to bring about substantive change. Organized communities have been better placed to defend their rights vis-à-vis bureaucrats and those who would undermine the reform effort. Finally, what actually gets implemented and the benefits that communities derive from the reform are a combination of struggle and opportunity. The take-away message for this author was of the importance of organized communities linked in their own organizations or networks that are able to take advantage of the opportunities presented. Herein lies the very important contribution of this book to the literature on social movements, agrarian change and the role of forests in poverty alleviation.

Carmen Diana Deere
(USA, July 2009)
Professor of Food and Resource Economics and
Latin American Studies, University of Florida

Preface

In recent years there has been a surge of research interest in forest management by local communities, and studies have documented the increasing share of forest resources that is owned by, or under management of, communities around the world. Less clear is the impact of such strengthened local forest rights. Do stronger community rights help local people to derive more, and more secure, benefits from forests? Do they strengthen incentives for sustainable forest management and protection? These were some of the questions surfacing in discussions in 2005 among the Center for International Forestry Research (CIFOR), Canada's International Development Research Centre (IDRC) and the Ford Foundation, when they began to consider how research might contribute to the newly established Rights and Resources Initiative (www.rightsandresources.org). The outcome of these discussions was a global research project of which the current book is a key output.

The project was led and coordinated by CIFOR and worked with a wide range of active research partners in ten countries, involving over 30 sites of differing size and characteristics. While not designed as a strictly comparative set of experiences, the broad variation in the sites enabled the project to examine forestry reform and community management across the globe in a large number of specific national and local contexts. To ensure a good degree of comparability a set of core concepts was formulated. Using a newly developed analytical tool (the 'tenure box'), the study examined the role of tenure rights across the sites in terms of 'LIFE' indicators: livelihoods, income, forest condition and equity.

In spite of the high degree of complexity that was found in all of the sites, a number of interesting findings have emerged from the study and are discussed at length in this book. The chapters were written to reflect the central cross-cutting themes identified in the research and address issues ranging from the nature of the reform, to co-management arrangements, the interface of new statutory with customary rights, relations of authority, social movements, forest regulations and markets. The outcomes suggest that positive results are more likely to be achieved if forestry reforms are fully implemented: interrupted or partial reforms do not work well. If reforms are to have a positive impact on equitable distribution of benefits, such equity concerns must be specifically built into the design. And while very complex trade-offs exist between livelihood improvement and forest protection, the study demonstrates that in a significant number of cases, community livelihoods can be enhanced through rights-based involvement in forest management without adverse environmental consequences. As the global community gears up towards major decisions on a future climate change mitigation regime (including possibly forest-related aspects), this book provides an important contribution to these debates.

Hein Mallee
(Singapore, July 2009)
International Development Research Centre (Canada)

Acknowledgements

This book would not have been possible without the collaboration of virtually hundreds of people from around the world. We are particularly indebted to the many community forest leaders and members who have helped us to understand the nature and effects of forest tenure reforms, some of whom endured hurricanes, food riots or violent conflict during the lifetime of the project and still supported our work. We hope that this research will contribute to securing their future livelihoods.

The study also depended on the support and quality fieldwork of our partner organizations and numerous researchers who participated in various phases of the research upon which this book is based. In Latin America, our partners in Nicaragua included the Universidad de las Regiones Autónomas de la Costa Caribe de Nicaragua (URACCAN) and Cooperativa de Profesionales Masangni; in Guatemala, Asociación de Comunidades Forestales de Petén (ACOFOP), Facultad Latinoamericana de Ciencias Sociales (FLACSO) and Facultad de Agronomía/ Universidad San Carlos (FAUSAC); in Bolivia, Centro de Estudios para el Desarrollo Laboral y Agrario (CEDLA); and in Brazil, the Assessoria Comunitaria e Ambiental (ARCA) and Laboratório Agroecológico da Transamazônica (LAET) of the Universidade Federal do Pará.

Researchers for the community and regional studies were Jadder Mendoza-Lewis, Ceferino Wilson White, Adonis Argüello, Arellys Barbeyto, Taymond Robins Lino, Onor Coleman Hendy, Constantino Romel, Marcos Williamson and Armando Argüello Salinas (Nicaragua); Iliana Monterroso, Silvel Elías, Juan Mendoza, Carlos Crasborn, Margarita Hurtado Paz y Paz, Rocío García, Aracely Arévalo and Blanca González (Guatemala); Marco Antonio Albornoz, Marco A. Toro Martínez and Roberto Ibarguen (Bolivia); and Westphalen Nunes, Patrícia Mourão, Rubem Lobo, Guilhermina Cayres, Ione Vieira, Ketiane Alves, Carla Rocha, José Antônio Herrera and Tarcísio Feitosa (Brazil). Deborah Barry coordinated the research at the global scale and in Africa and also participated in Guatemala; Anne Larson coordinated research in Central America; Peter Cronkleton, in Bolivia; and Pablo Pacheco, in Brazil and Bolivia.

In Africa, partners included Civic Response in Ghana and in Burkina Faso, the Panafrican Institute for Development (PAID), Poda Damas from the Forest Department in the Ministry of Environment and Kimse Ouedraogo, Director of the National School of Foresters. Research and technical support were provided by Emmanuel Marfo, Kyeretwie Opoku, Rebecca Teiko Dottey, Richard Adjei, Mr. A. Baah, Mr. Kyei, Mr. Wallas, Mr. Kyei Yamoah, Osei Tutu and Mr. Damte (Ghana); Phil René Oyono, Serge S. Kombo, Martin B. Biyong, S. Mbuh, J. Monda, Z. Ngbamine, D. Maschouer, R. Meva'a, L. Tatah Shulika, A. Akeh, S. Evina, S. Eva Meyo, B. Tchoumba, J. Abbé, P. Paah, J. Tsana, C. Ndikumangenge and A. Minsouma (Cameroon); and Bocar Kante, Amadou Diop, Mody Kone, Mansour N'Diaye, Alfred Ouedraogo and the late Manadou Traore (Burkina Faso).

Partner organizations in Asia included, in Nepal, ForestAction and the Federation of Community Forestry Users, Nepal (FECOFUN); in the Philippines, the University of the Philippines, Los Baños; in India, a PhD student from the University of Cambridge; in Laos, the Bangkok-based Regional Community Forestry Training Centre (RECOFTC). Researchers included Juan M. Pulhin, Josefina T. Dizon, Rex Victor O. Cruz, Dixon T. Gevaña, Mark Anthony M. Ramirez, Hanna Leen Capinpin and Chandellayne G. Cantre (Philippines); Naya Sharma Paudel and Mani Ram Banjade (Nepal); Sushil Saigal (India); and Richard Hackman and Robert Oberndorf (Laos). Ganga Ram Dahal coordinated the Asia research.

Krishna Adhikari prepared a very useful regional literature review on forest tenure as well as a summary of information across the sites in Asia, and Anne Larson and Deborah Barry prepared a literature review on Latin America; similarly, Jerome Attaia Ndze and Phil René Oyono prepared an annotated bibliography for Central and West Africa. Cyprain Jum prepared helpful databases and summaries of information across the sites in Africa and Phil René Oyono prepared this information for Cameroon. Monico Schagen and Yulia Siagian prepared a review report on decentralization and tenure reform in Indonesia. This collection also represents a contribution to the IUFRO (International Union of Forestry Research Organizations) Task Force on Improving the Lives of People in forests.

The editors would like to thank all the authors for their ongoing contributions throughout this process, including travelling around the world to participate in two workshops to discuss, analyse and write up and share results – hence every chapter was truly a collaboration beyond the specific list of authors. Reviewers' comments on each chapter also proved invaluable: thanks go to Grenville Barnes, David Bray, Ashwini Chhatre, Chimère Diaw, Chip Fay, Bob Fisher, David Kaimowitz, Hein Mallee, Ruth Meinzen-Dick, Nick Menzies, Moira Moeliono, Augusta Molnar, Thomas Sikor, Anthony Stocks, William Sunderlin, Tim Synnott and Peter Taylor. We want to thank William Sunderlin for helping with the tenure data and Mohammad Agus Salim for the map. We are particularly grateful to Rahayu Koesnadi for her patient and unwavering administrative support throughout this project and Sally Atwater for her excellent editing of the final manuscript.

We also want to thank participants and supporters of the Rights and Resources Initiative, including Jeff Campbell, Doris Capistrano, Hein Mallee, Ujjwal Pradhan and Andy White, for their help in the early design stages of this research project, and Doris Capistrano and Carol Colfer, and in the later stages Elena Petkova, who provided substantive, administrative and logistical support from CIFOR headquarters.

Finally, we are grateful for the support of the International Development Research Centre (IDRC) and the Ford Foundation for the generous financial support that made this research possible in all three regions, as well PROFOR/World Bank for its support in Latin America specifically. We conclude with a special thanks to Hein Mallee of IDRC for maintaining his faith in and support for this project from beginning to end, even through its difficult moments.

Part I
Introduction

1

Tenure Change in the Global South

Anne M. Larson, Deborah Barry and Ganga Ram Dahal

In Asia, Africa and Latin America, governments are granting new tenure rights to communities living in and around forests. An important shift in forest tenure has occurred since 1985, with at least 200 million hectares (ha) of forest recognized or legally transferred to communities and indigenous people (White and Martin, 2002). In a study of 25 of the 30 most-forested countries, Sunderlin et al (2008) found that 15 countries had experienced an increase in land designated for and/or owned by communities since 2002 alone. Today, then, 74.3 per cent of the global forest estate is owned and administered by governments; 2.3 per cent is owned by governments but designated for use by communities; 9.1 per cent is owned by communities and the remaining 14.2 per cent is owned by individuals and firms (Sunderlin et al, 2008). The percentage of forests in the hands of communities[1] in the developing world alone is much higher: 22 per cent in 2002 and 27 per cent in 2008 (Hatcher, personal communication, based on data from Sunderlin et al, 2008).

The change in forest tenure constitutes a kind of 'forest reform' (Pacheco et al, 2008a; Taylor et al, 2007), comparable to the widespread agrarian reforms of the mid-20th century. The current reforms are due to the growing recognition of rights and benefits belonging to people living in and around forests. They may originate as much 'from above' as 'from below', with forces driving and shaping reforms emerging from communities and social movements, international donors or the state.

This book explores the nature, goals and results of such reforms in practice. It is based on research at more than 30 sites in 10 countries that have all promoted, in some way, greater local rights to forests. The countries are in Asia (India, Nepal and the Philippines), Africa (Burkina Faso, Cameroon and Ghana) and Latin America (Bolivia, Brazil, Guatemala and Nicaragua). Less intensive research was also conducted in Lao People's Democratic Republic (Lao-PDR) and work in Indonesia has already been well described in CIFOR

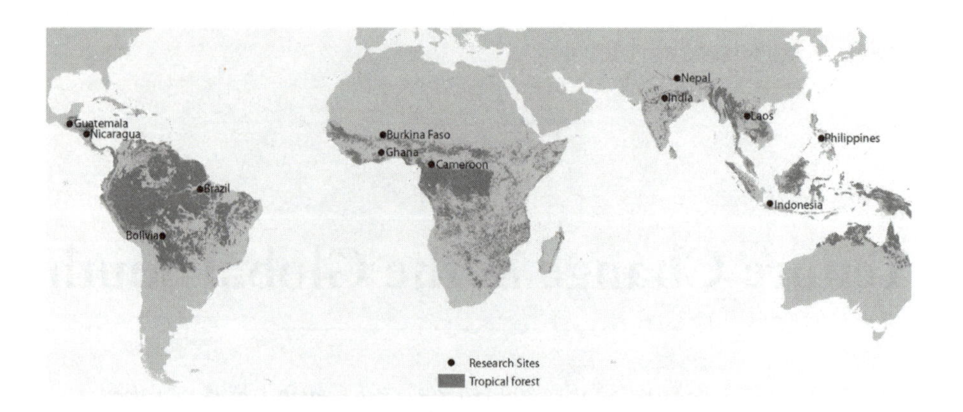

Figure 1.1 *Map of the research sites*

publications (Figure 1.1). Literature reviews were conducted in all three world regions to help ground the analysis of field research in historical and cultural contexts.

Forest tenure reforms range from the titling of vast territories to indigenous communities, to the granting of small land areas for forest regeneration or the right to a share in timber revenues. New statutory rights do not automatically result in rights in practice, however, nor do local rights necessarily lead to improvements in livelihoods or forest condition. To understand the meaning of new rights it is important to know what rights people held previously, particularly *de facto,* or customary, rights, since formal statutory rights may place new restrictions on communities. Because people held rights previously in most cases, it is often more appropriate to refer to the recognition or formalization of rights, rather than the transfer of rights.

In most countries, the research was undertaken specifically with local partners, sometimes the new rights holders themselves, who would be able to take advantage of the process and its results to promote community rights, effective forest management and livelihood opportunities. Preliminary research found that institutional weaknesses and policy distortions have limited the impacts of change. Hence the project was designed to generate information not only for academic analysis but also to promote empowerment and engage in effective policy dialogue with non-governmental organizations (NGOs), donors and governments.

To reflect our findings, this book takes an ambitious approach. Rather than present the results of our research through chapters on individual case studies, we address cross-cutting issues that we believe capture the essence of the reforms: the challenges they face and the opportunities they unlock. Each topic constitutes a central aspect of forest governance and builds not only on the case studies but also on the existing literature and the experience of each author.

This introduction places the research in context. It begins with a brief discussion of forest tenure reform in light of community rights and forest conservation and a short introduction to each region, leaving the main discussion of other governance issues to the chapters. The next section introduces the research project itself, including the goals, methods and models of tenure reform studied. This is followed by the definition of the concepts used in the study and in this book. The subsequent section introduces the chapters and the final section reminds us of the global context in which the tenure reform is playing out.

Why forest tenure reform?

The literature on forests and on conservation is replete with cases of rural communities whose livelihoods have been affected by state policies or the intrusion of outsiders into 'their' forests. These include state-authorized forest concessions (e.g. Anaya and Grossman, 2002), forest classification schemes that prohibit community use (e.g. Peluso, 1992), mining and petroleum concessions (e.g. Oyono, et al, 2006; Kimerling, 1991; Lynch and Harwell, 2002), evictions from, or severe limitations on their livelihood activities in, parks or protected areas (e.g. Dowie, 2005; Spierenburg et al, 2008; Cernea 1997, 2006; Brockington and Igoe, 2006) and colonization or invasions by farmers and ranchers (e.g. Schmink and Wood, 1984; Colfer et al, 1997; Fulcher, 1982; Baird and Shoemaker, 2005). In many cases, these forests, historically, had been used and managed by communities themselves.

Colonial policies justified the centralization of forests based on 'scientific forestry' principles (see Chapter 7). On the one hand, forests were seen as public goods and strategic resources that needed both protection and 'rational use' in order to provide both goods and income for the future. On the other, however, their exploitation often favoured elite interests over others. For example, in Ghana,

> *...before 1924, natives held [forest] concessions and sold wood upon the same basis as Europeans. But the competition became so keen...that in a 1924 administrative order, the government declared that a native could not cut and sell wood except for his own use without making a deposit with the government of twenty-five hundred francs – a prohibitive sum.* (Buell, 1928, p256, cited in Larson and Ribot, 2007)

In addition, explicitly discriminatory policies have also sometimes been accompanied by corruption, rent seeking and the creation of patronage networks by government officials – patterns that continue to this day (Larson and Ribot, 2007; Sunderlin et al, 2008).

From purely a rights perspective, there is little room for doubt that many communities living in forests today deserve a better deal. Numerous grassroots organizations and movements around the world have spoken out to demand rights to forests. Latin American indigenous movements, in particular, have sought, in some cases successfully, to regain traditional rights over their historic territories and forests. At the same time, research has begun to examine the effect of forests on vulnerability (e.g. Hobley, 2007) or their potential role in poverty alleviation (Sunderlin et al, 2005).

But what about forest conservation? Sayer et al (2008, p3) write, 'The harsh reality for conservation is that, for most local people, conversion to agriculture or to industrial estate crops provides a faster route out of poverty than either local forest management or total protection.' There is no guarantee that local people will conserve forests if they have more, or more secure, rights, though the central tenet – that secure rights permit longer-term horizons and greater interest in sustainability – appears to hold. In some cases, however, converting forests to other uses will bring greater livelihood benefits and may even be sustainable over the long term (Tacconi, 2007a). In others, more secure tenure rights have clearly improved forest management (Sayer et al, 2008).

What will work best for conservation depends on the causes of deforestation and degradation. In some cases the state itself promotes logging, clear-cutting and conversion to industrial crops, as in Indonesia, which has one of the highest rates of deforestation in the world (FAO, 2005). In other cases, multiple interests in forests and forestlands have led to invasion, colonization and conversion. Our research finds that where communities have demanded tenure rights, a common reason is outside encroachment on their land. In Latin America, there is substantial overlap between standing forest and indigenous communities (see www.raisg.socioambiental.org) and land invasions by external actors are a leading cause of deforestation (Geist and Lambin, 2002; Stocks et al, 2007).[2] Securing community tenure rights – and, in particular, defending their exclusion right – could thus be essential for conservation.

The fear that forest conversion and degradation will continue apace under community tenure has served to justify not only state forest regulation but also sometimes heavy restrictions on forest use accompanying forest tenure reforms. As discussed in the next chapter, conservation interests continue to propose solutions that still sometimes remove people from protected areas, but many people believe that governments have failed to maintain forests and that conservation cannot work if local people don't 'buy in'.

The use of land and forest resources has played out differently in the three main regions of the developing world and set the stage for reforms under different sets of parameters. For example, population densities in Asia contrast with the vast expanses of forest per household of the lowland forests of the Amazon. The nature of colonialism was different in Latin America and ended far longer ago. Ongoing wars and population movements mark present-day Africa. At the same time, the historical centralization of forests – and denial of community rights – is common to all, as are ongoing deforestation and forest degradation. Remote forests have remained largely under customary practices and are somewhat protected from outside pressure. All three regions are experimenting with granting new forest tenure rights to local communities and each will be considered briefly in turn.

In Asia, the failures of centralized ownership and management of forests led to a rethinking of forest management and tenure policies in many countries as early as the 1970s. China, Nepal, Thailand and the Philippines banned timber exports; several countries placed heavy regulations on industrial concessions (Adhikari, 2007). At times, timber concessions were cancelled or not renewed, sometimes causing a shift to plantation forestry, as in Indonesia. The emphasis on wood production shifted to plantations and in several countries up to 90 per cent of raw material is now supplied from trees outside natural forests (Enters et al, 2003).

Policy-makers in India and Nepal observed that denying local communities access and management rights to forests worked as a disincentive, exacerbating forest degradation, conflicts and poverty. India, Nepal and the Philippines led Asia in introducing policies aimed at formally involving local communities in forest management; other countries (e.g. Laos) followed. Policies in Bhutan, Cambodia, China, Sri Lanka, Thailand and Vietnam are still in their formative stages (Gilmour et al, 2004) but have emphasized the recovery of degraded forests. Countries with ample forest resources demonstrate patterns different from those with either seriously degraded or less valuable forests. In particular, Asian governments have been less likely to recognize local rights if the country has rich forest resources.

In Africa, statutory forest tenure is characterized by almost exclusive public administration: 98 per cent of forests are under the formal control and management of government authority (see Figure 1.2). Even in 'state-owned' forests, however, customary authorities, law and practices (as in

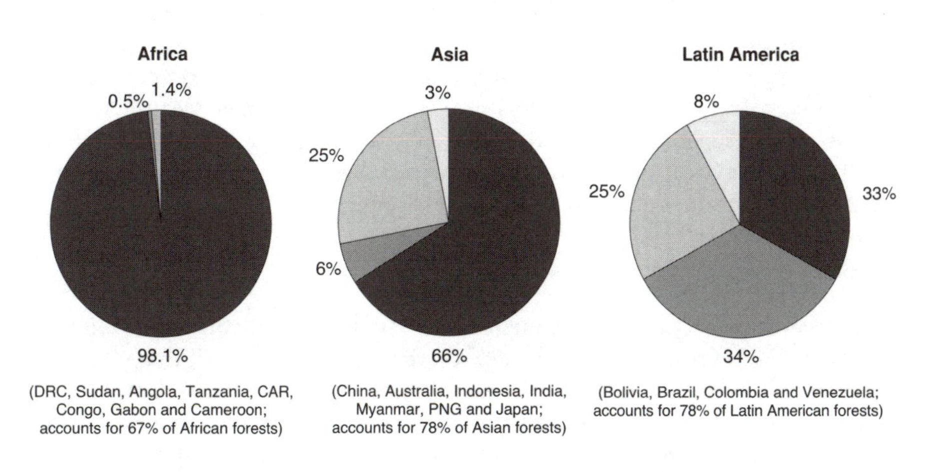

Source: RRI (2009)

Figure 1.2 *Forest tenure distribution among three world regions, 2008*

Asia) play a significant role in local governance and resource access. Forest policies, specifically, have been reformed in the vast majority of countries since 1990, and in a single decade, 'more than 30 countries launched at least one significant ground initiative towards community participation in local forest management' (Alden Wily, 2004). Nevertheless, governments generally retain most of the decision-making power for forest management either through exclusive control of forests or by granting only non-commercial user rights to satisfy the needs of local people for forest products. Forestry sector reforms have been driven primarily by decentralization policies (see Chapter 2), as well as some community forestry initiatives.

Of the three regions, Latin America has the smallest portion of land under government ownership and management (33 per cent) and the largest in the hands of private individuals and firms (34 per cent). The area owned by communities, 25 per cent, is similar to Asia, and an additional 8 per cent is public land designated for community use (see Figure 1.2). In this region, Mexico is at the forefront of community forestry. Agrarian policies dating to the Mexican revolution and granting land rights in subsequent waves over much of the 20th century laid the foundation for community rights to forests, and 'a vigorous community forestry sector emerged beginning in the 1970s' (Bray et al, 2006). A community forestry pilot project, known as the *Plan Piloto Forestal*, was launched in Quintana Roo in the early 1980s (Salas, 1995) and by the 1990s community forestry was widespread in other parts of Mexico as well (Bray et al, 2005).

In most of Latin America, recent changes in tenure were the result of grassroots struggles. Indigenous peoples have sought the recognition of their ancestral domains, as have numerous non-indigenous populations, such as rubber tappers in Brazil. These groups have historically lived in and maintained forests despite both state intervention and pressures from competing land claimants: sometimes poor, landless peasants, and sometimes wealthier, agro-industrial or logging interests. Reforms in Latin America are more likely to involve the demarcation and titling of large indigenous territories, with large expanses of land for relatively few people (Stocks, 2005).

Research sites

Our research project emerged from interest in understanding the tenure changes that were taking place around the globe, with the goal of catalysing efforts to advance local control and management of forests. This involved promoting research and action across multiple scales, as well as various adaptations of the methods and choice of research sites across countries based on the different types of reforms and on the more action- and policy-oriented goals. Scoping allowed us to scout out the most appropriate locations for both research and action goals, as well as to find experienced and knowledgeable partners.

The central analysis took place in 'research sites', usually multiple villages, where a change in tenure or resource rights had occurred, or was about to take place. Sites were chosen not only to explore tenure changes but also based on apparent opportunity to deepen rights or affect policy decisions. In Latin America, the research sites tended to be large areas of 1 million to 2 million ha because reforms were based in specific regions around large territories. In Africa and Asia, the sites were much smaller but still involved multiple villages or communities organized around a specific forest.

The research involved scaling both downwards and upwards from each site. At the more local scale, it was aimed at examining socially and economically differentiated access to forest resources, institutional processes and mechanisms for sharing benefits within and among communities under tenure reform. In addition to providing the basis for collecting specific data, our work was aimed at informing strategies and processes for enhancing the rights and livelihood security of vulnerable groups, as well as increasing institutional capacities and leadership skills within grassroots organizations, federations and movements. This was intended to enable these actors better to represent and articulate the interests and priorities of their local constituencies, especially vulnerable groups within them, and to engage effectively with decentralized structures and policy-making processes.

At the larger sub-national and national scale, action-focused research built on more general findings to inform dialogues between governments and civil society organizations and to identify constraints and opportunities for linking pro-poor forest management to decentralized, as well as central government, planning processes.

We chose the sites and research communities that would provide the best understanding of tenure reforms in each national context with the resources at hand. Hence, depending on the nature of the reform, our field sites were typical cases or particularly interesting or exceptional experiences. Within a single country, the sites chosen may represent different types of reform, types of forest, forest classifications or types of market engagement. These cases were then analysed in relation to research into the broader regional and national context. We considered contextualization a critical feature of our approach. The cases in this book, then, sometimes refer to national policies, sometimes to a particular model of change and sometimes to the experience of a particular community. Table 1.1 lists the countries, regions and sub-regions and the 'communities' (defined below) studied, as well as a brief description of the model of forest tenure reform.

Some aspects of the research methods also varied from country to country and even from site to site. In almost all cases, partner organizations were identified to spearhead the research and methods at the site scale ranged from more participatory to more conventional, depending on available capacities. In all cases, lead researchers at the country or sub-country regional scale – almost always developing country nationals and always with extensive experience in the regions studied – were charged with oversight of the site-level research, guaranteeing effective analysis of the findings in light of the particular research context.

All the research was carried out using the same set of central questions, theoretical and background readings, hypotheses and definitions of terms. In

Table 1.1 *Research sites and tenure models studied*

Country	Region	Community	Tenure model
Bolivia	Guarayos	Santa María de Yotau	Communities within indigenous territory being demarcated and titled
		Cururú	
	Northern Amazon (Pando)	Turi Carretera	Agro-extractive communities being demarcated and titled
		San Jorge	
Brazil	Porto de Moz	Turu	Agro-extractive communities bordering agro-extractive reserve (RESEX)
		Taperu	
	Trans-Amazon	Dispensa I	Colonist communities
		Pontal	
Guatemala	Petén	Carmelita	25-year community forest concession (community living inside concession)
		Arbol Verde	25-year community forest concession (members from several communities living outside concession)
	Highlands	Chancol	Highland communal forests (multiple community, single title, community owned)
		Mogotillos	Highland communal forests (local government owned)
		Chichim	
		Estancia	
Nicaragua	RAAN	Tasba Raya	Indigenous territories being demarcated and titled
		Layasiksa	
Burkina Faso		Goada Forest	Local association: management for regeneration
		Nakambé	Concession: fuel wood management (classified forest, central government domain)
		To	Concession: fuel wood management (non-classified forest, local government domain)
		Comoé-Léraba	Concession: forest and wildlife reserve
Cameroon	Lomié/Dja	AVILSO	Community forests
		Medjoh	
	Mount Cameroon	Bimbia–Bonadikombo	
	Northwest Cameroon	Oku	
	South Cameroon	UDEFCO	
		Kienké–Sud	Forest revenue sharing (logging concession to company)

Ghana	National		Benefit sharing from logging
	Assin Fosu Forest District (Adwenase Community Forest)	Assin Akropong	Collaborative forest management: 'dedicated forest' with management plan, for protection
		Subinso 1	
		Subinso 2	
	Afram Headwaters Forest Reserve	Asempanaye	Modified Taungya System (tree planting, community and individual farmers share future timber revenue)
		Ada Nkwanta	
		Kwapanin	
India	Ajhmer, Rajhastan	Kumhariya	25-year renewable lease for tree grower cooperatives for fuel and fodder to recover wastelands
		Nathoothala	
		Khoda Ganesh	
Nepal	Nawalparasi, Terai (lowlands)	Sunderi CFUG	Community forests with approved operational plans
	Lalitpur (hills, periurban)	Patle CFUG	
	Baglung (hills, rural)	Sanghukhola Ratopahara CFUG	
	Dolakha (high-altitude hills)	Suspa CFUG	
Philippines	Nueva Vizcaya, Region 2	Kalahan Education Foundation	Certificate of ancestral domain with community based forest management
		Banila Community based Cooperative Project	Community-based forest management
		Barobbob Ecological Socio-Economic Project	Co-management agreement with local government (local occupation rights for 25 years, renewable for 25 years)
	Compostela, Mindenao	Nagan-Panansalan-Pagsabangan Forest Resource Development Cooperative	Community-based forest management

the end, we have collected and analysed a wealth of multiscalar information obtained through diverse entities and methods and covering a broad range of types of reform in multiple contexts.

Themes and concepts

The research questions were organized around five themes: tenure change, local organization, the role of regulations, engagement with markets and outcomes for livelihoods, forests and equity. These can be summarized as follows:

- What is the effect of tenure change on community rights to access and decision-making regarding forests?
- How do the regulatory framework, markets and local organization affect these rights in practice?
- What are the interactions among these variables or spheres?
- What is the effect of each on outcomes?

Understanding the tenure change and its effects on existing rights and practices were taken as the points of departure. Local organization, regulations and markets were primarily conceived as mediating variables that would permit or inhibit better outcomes. Outcomes were measured based on a combination of variables summarized as livelihoods, forest condition and equity (see Chapter 9). This section presents our understanding of the terms used throughout the research and this book.

Tenure rights are conceived of as a bundle of rights, ranging from access and use rights to management, exclusion and alienation (see Schlager and Ostrom, 1992). *Access* refers simply to the right to enter the area. *Use,* or *withdrawal, rights* refers to the right to obtain resources, such as timber, firewood or other forest products, and remove them from the forest. *Management* refers to 'the right to regulate internal use patterns or transform the resource' (Agrawal and Ostrom, 2001, p489), which could include tree planting, timber management or conversion to agriculture. *Exclusion* is the right to decide who can use the resource and who is prevented from doing so. *Alienation* is usually understood as the sale or lease of the land, which also includes the sale of these other rights. The last three rights are seen as decision-making rights and are, therefore, particularly significant for tenure reforms.

Resource tenure consists of the social relations and institutions governing access to and use of land and natural resources (von Benda-Beckman et al, 2006). *Forest tenure,* then, is concerned about who owns forestland and who uses, manages and makes decisions about forest resources. Forest tenure determines who is allowed to use which resources, in what way, for how long and under what conditions, as well as who is entitled to transfer rights to others and how. Different elements of the bundle of rights may be shared or divided in a number of ways and among stakeholders; in addition, trees themselves may be subject to multiple tenure rights (Fortmann, 1987).

The bundle is also likely to include a combination of rights that are defined by statutory law (*de jure*) and rights that are defined locally, through *de facto* or customary institutions. A *de jure* right concerns a set of rules established and protected by the state (e.g. registered land titles, concession contracts, the forestry law and regulations). *De facto* rights are patterns of interaction established outside the formal realm of law. They include *customary rights*, a set of codified community rules and regulations inherited from ancestors and accepted, reinterpreted and enforced by the community, and which may or may not be recognized by the state.

Tenure reform is the legal reform of tenure rights (Pulhin and Dizon, 2003). *Forest tenure reform* is different from land reform: the latter entails redistribution of landholding and changes in the agrarian structure, whereas the former is a change of one or more rights regarding forest resource and forest land management (Bruce, 1998). Forest tenure reform usually involves granting rights to people already living in or near forests and using forest resources (see Chapter 2).

Property here refers to real estate, whereas *tenure* refers to the way rights are administered, though property and tenure are often used interchangeably. Property is usually classified as either private or state. Communal tenure systems and common property exist on either state or privately owned (communal) lands. Most of the world's forests are formally state owned; forest tenure reforms usually give forests to collectives under *communal tenure* regimes as *communal property*. Within a communal land area, there may be both common and individual properties and decisions may be made individually or collectively, but the holder of the right is still the collective. To work through this web of definitions and relations, we adopted the 'tenure box' from Meinzen-Dick (2006), which enables one to tease out the bundle of rights on the one hand and the rights holders on the other (see Chapter 3).

Tenure security is the degree to which an individual or group believes its relationship to land or other resources is safe, rather than in jeopardy (Poffenberger, 1990). We do not assume that any particular configuration of rights, such as a land title, constitutes security (Ellsworth, 2002).

The *community* was the basic unit of analysis across all sites. *Community* does not necessarily refer to a group of people who live in a single village but rather is defined as those who share a common interest or purpose in a particular forest and share common resources. Hence the resident-based community (or village) may overlap with the community of interest or be a subset of it, or vice versa. There may also be local 'communities' embedded in larger communities.

Community forestry is understood broadly as a common property resource management approach with characteristics and institutional innovations devised by local people (Chapagain et al, 1999) to organize and exercise their rights for the use and management of a forest area for the supply of forest products. Though it sometimes refers to a type of project promoted by the state or donors, it does not refer only to such projects.

Content of chapters

Through the comparison of selected cases, the chapters explore the nature of forest reform, the extent and meaning of rights transferred or recognized, the role of authority and of citizens' networks in forest governance, opportunities and obstacles associated with government regulations and markets for forest products, and the outcomes for livelihoods, forest condition and equity. The ten chapters, including this introduction, are organized into five sections.

Chapter 2, by Deborah Barry, Anne Larson and Carol Colfer, completes Part I of this book by exploring the nature, origins and important global processes shaping forest tenure reforms. These reforms, initiated more aggressively since the mid-1980s, appear sufficiently widespread to constitute a global trend. Although the findings are grounded in the case studies and countries studied, a step back to the broader continental and global scale provides insights into the larger forces shaping the reforms and allows for their preliminary characterization.

The authors find this trend significantly different from previous agrarian reforms, when forests and cleared lands were transferred to peasant farmers for agricultural purposes. Rather, these new reforms are being driven and shaped by local claims for tenure rights recognition, the global concern for biodiversity conservation and the promotion of democratic decentralization. They aim to accomplish three goals: addressing claims to historic rights, improving local well-being and achieving forest conservation. The chapter argues that policy-makers need to understand better the nature of this tenure reform and radically adjust their goals, institutions and regulations for the implementation of what could become one of the most important global efforts to thwart growing rural poverty and mitigate the effects of climate change.

Part II analyses two issues in the process of transferring or recognizing community tenure rights: the nature of management rights devolved or withheld and the interface of statutory and customary rights. Chapter 3, by Peter Cronkleton, Deborah Barry, Juan Pulhin and Sushil Saigal, draws on case studies from Guatemala, Bolivia, the Philippines and India to examine the issue of management rights. The authors present these four cases of reform using the tenure box, mentioned above, to examine the type and characteristics of rights devolved and how this influences community forestry models and the benefits received by community-level participants.

The devolution of forest tenure rights to local stakeholders around the world has produced a variety of community forestry models. Some kind of co-management is usually involved such that the state maintains ownership and control over forest resources, either authorizing use of state lands or requiring forest users on non-state property to operate under government supervision and its norms. These arrangements recognize some existing resource uses embedded in local livelihoods and customary practice but also introduce new rules and techniques and restrict certain previous behaviours. Such arrangements are mainly, at least ostensibly, intended to promote greater sustainability and equitable use, but they can also introduce disincentives and distortions and severely limit local decision-making power. This not only attenuates tenure rights but may also undermine previously effective local management institutions and reduce livelihood benefits.

Recognition of tenure rights for communities already living in forests almost always encounters existing *de facto* or customary arrangements. Chapter

4, by Emmanuel Marfo, Carol Colfer, Bocar Kante and Silvel Elías, analyses the interface between statutory and customary land laws and rights. The authors use experiences in four countries with strong traditions of customary rights – Ghana, Indonesia, Burkina Faso and Guatemala – to examine the extent, models and forms of acceptance or recognition of customary systems by formal law under the forest reforms. Security of tenure is fundamental to good governance and poverty alleviation, but secure tenure is bedevilled by overlapping legalities that impose multiple, simultaneous systems of rights, each with its own source of legitimacy. Disregard for this complexity creates unexpected outcomes and can fuel conflict.

In almost all the cases studied in this book, tenure rights are defined by both statutory and customary laws. However, there is debate as to whether one legal system should be considered superior, whether there should be a sharing of legitimacy between legal systems, or whether legal pluralism should allow for different tenure systems to coexist simultaneously. Among the options is a shift from legal pluralism to legal integration, combining the strengths of both customary and statutory laws. To make such an endeavour possible, it is important to document how statutory and customary laws within specific socio-political settings have coexisted and played out. The chapter examines, in each of the four countries, the extent to which statutory law has accommodated (recognized) or subverted customary systems of tenure and how it has done so, through different models of recognition.

Part III turns to two aspects of 'local' forest governance institutions: local 'authorities' and social movements. Chapter 5, by Anne Larson, Peter Cronkleton, Juan Pulhin and Emmanuel Marfo, explores the configuration of authority relations in forest tenure reforms in three indigenous territories (in Nicaragua, Bolivia and the Philippines) and in Ghana. When rights are granted or formalized to a 'community', a new or existing institution is often designated to represent this collective. That institution is then likely to shape the exercise of the new rights on the ground, based on its nature (whether it is representative or accountable, for example) and domain (the powers it holds). This can be particularly problematic for indigenous territories that did not previously have a common governance structure.

Chapter 5 looks specifically at how the recognition of community rights by central governments tends to lead to political contestation over authority. The politics of authority takes different forms in the four cases examined and the findings suggest that this issue should receive much greater attention in reforms. These contestations often lead to conflict and the breakdown or manipulation of authority relations, but they may also allow new configurations of effective, representative and accountable authority to emerge.

Chapter 6, by Naya Paudel, Iliana Monterroso and Peter Cronkleton, examines the central role of a fairly new kind of social movement that has so far received little attention in the literature: networks of community members and organizations that scale up for collective action to defend rights and expand opportunities for community forest management. Three such organizations are discussed: the Federation of Community Forest Users, Nepal (FECOFUN), the

Association of Forest Communities of Petén (ACOFOP) in Guatemala and the Brazil Nut Producers' Cooperative (COINACAPA) of Bolivia.

The chapter synthesizes the conditions for the emergence and evolution of these organizations, their institutional dynamics and strategies of resource mobilization, as well as their effects on resource tenure and livelihoods. The first two organizations have played a central role in obtaining, defending and deepening the rights of their members, while the last emerged primarily to improve members' market position and incomes. The chapter argues that networks have increased local agency in the tenure reform process, improved the institutional and technical capacity of communities and greatly enhanced the abilities of communities to influence public opinion, policy and the regulatory framework governing their forest rights.

Part IV turns to a discussion of regulations affecting reforms and both timber and non-timber product markets. Even where substantial new and secure rights have been granted, government regulations – and associated transaction costs – may prevent community access to forest products and markets and thus doom the livelihoods potential of reforms. Chapter 7, by Juan Pulhin, Anne Larson and Pablo Pacheco, reviews cases involving three kinds of regulations. First, governments often limit the kind of forests available for communities, giving them wasteland or degraded forests for tree planting or protection, rather than high-value forests that could generate significant income, and reserve the best areas for the state (which in turn grants them in concession to industry). The second type addresses limitations on resource use in conservation areas. The third set of rules refers specifically to forestry regulations, such as permits for logging.

The authors draw on experiences in India, Brazil, Nepal, the Philippines, Guatemala and Cameroon. Though some rules are surely needed to conserve forests, regulations are often unrealistic and unenforceable and/or an incentive for graft and corruption. The power of self-perpetuating bureaucracies may need to be broken to create new regulatory frameworks that are more responsive to people and relevant to diverse local realities.

Communities engage with markets both formally and informally, and Chapter 8, by Pablo Pacheco and Naya Paudel, explores the associated challenges and opportunities. Forms of market engagement are shaped both by community capacities and by the degree of development of the markets themselves. The authors take issue with extreme views that markets are either a panacea for communities or simply a way for outside actors to extract economic rents. The cases examined include two community logging enterprises (in Nicaragua and Bolivia); two communities engaged in the sale of non-timber forest products (Bolivia and Nepal); two situations in which smallholders make individual rather than collective decisions, particularly for timber sales, and often operate outside the law (from Bolivia and Brazil); and one case in which communities log for timber but are allowed to sell only to their members (Nepal).

Each market is different, and the regulatory framework within which communities operate affects the cost–benefit analysis of their marketing choices. The different market conditions suggest broad room for policy action:

tenure reforms should not only focus on building community capacities for market engagement but should also address specific market conditions under which communities and smallholders operate.

Part V presents the conclusions. In Chapter 9, Ganga Ram Dahal, Anne Larson and Pablo Pacheco discuss the outcomes of reforms by examining changes in livelihoods (including income), forest condition and equity across the sites. Though numerous dynamic processes affect outcomes, this chapter attempts to isolate the effects of tenure reforms, based on the assumption that tenure rights and security shape the decisions that local people make regarding forest use and management. Hence the first task is to determine the extent to which rights have actually increased and are secure, then to assess outcomes.

The chapter finds that most of the reforms resulted in some improvement in livelihoods. These may be quite small or counterbalanced with (sometimes temporary) hardships suffered by certain actors, such as seasonal pastoralists or poorer members of the community. Much larger income gains are associated with larger and higher-quality forests and community logging enterprises, but these benefits are not always possible in the small forests often granted to communities in Asia; nor are they necessarily better, because of the ways in which these projects are sometimes implemented.

Forest conditions most clearly improved when the reform specifically involved tree planting or improving degraded forests, and some cases with little noted change showed much less degradation than nearby forests in the same region. Declines in forest conditions appeared mainly where there were competing demands on forests, such as proximity to colonization areas and other large-scale dynamics beyond the control of the communities, suggesting the need for effective governance at larger scales as well. With regard to trade-offs between livelihoods and forest condition, several cases demonstrated livelihood improvements without declines in forest condition.

Positive outcomes in equity appear to depend on specific, dedicated efforts to address sources of inequity.

In Chapter 10, Anne Larson, Deborah Barry and Ganga Ram Dahal conclude the book by returning to the important findings and discussing cross-cutting issues and concerns raised by the research. These issues are discussed in light of emerging global challenges and opportunities regarding community rights and the future of forests, particularly global climate change. The authors close with a reflection on the future of forest tenure reforms.

Moving forward

Despite the enormous differences in the historic processes of defining land and forest resource rights in the countries studied, we have come to an initial understanding of the particular characteristics of incipient forest tenure reform. Greater clarity on the nature of forest tenure, how it is being shaped and how it could be promoted more consciously is essential for the success of any attempt at conserving the world's forests and improving the lives of its poorest peoples.

This task is of greater importance at a time when the forests of the world, particularly developing countries' tropical and dry forests, have become an important arena of global debates and plans to ward off the imminent perils of a rapidly changing global climate. The issue of rights to forest resources underlies the entire host of decisions being made in relation to forests. Mitigation schemes referred to as reducing emissions from deforestation and forest degradation (REDD), together with emerging markets for forest carbon credits (capture and storage of carbon), are introducing yet another dimension of rights: who will own the carbon? Adding a 'layer' of international rights to carbon on the existing web of forest tenure rights will require taking a closer and harder look at current trends, if people living in forests are, on one hand to avoid harm and, on the other, to benefit from proposed solutions. A reading of the changes in forest tenure from a rights perspective, including the perception of local forest dwellers, is paramount for working through the emerging contradictions and tensions.

In this light we now turn to the next chapter, which analyses some of the most important global processes motivating and shaping tenure changes, often with conflicting goals and inappropriate or overlapping institutions for their implementation. Understanding the forces defining this process of transition in forest tenure can help elucidate the nature of the challenge we face for bringing about just, coherent and workable change.

Notes

1. Either as owners, or lands that they have been granted the rights to manage.
2. The more common dynamic in Latin America that involves conversion of forest to pasture should not be confused with the debates regarding shifting cultivation (e.g. Angelsen, 1995).

2
Forest Tenure Reform: An Orphan with Only Uncles

Deborah Barry, Anne M. Larson and Carol J. Pierce Colfer

Significant tenure reforms in public forestlands have taken place over the past 20 years worldwide, but particularly in Latin America. These reforms, initiated more aggressively since the mid-1980s, appear widespread enough to constitute an important global trend. They also present an opportunity to advance the recognition of human rights and two critical values: providing benefits for poor forest dwellers and conserving forests for environmental reasons. Such tenure reforms may be essential to successful forestland governance.

Tenure and property rights are among the defining institutions of a society and are deeply embedded in regional and national history and local culture. In each region studied, the imprint of the major colonial powers (Spain, Portugal, Britain, The Netherlands and France) left very different land and forest regimes and legal systems. These systems also defined the rules for exclusion and the meaning of community, embedding a wide variety of relationships around land and forest resources (e.g. settler colonies, rule through 'traditional' chiefs, plantation regimes, peonage and slavery). The current tenure reforms can be properly understood only within these larger contexts. The findings in this book result from mapping these recent changes at the community level, understanding who receives what rights and recording and analysing how regulatory frameworks and market challenges work for or against community interests.

A significant challenge is seeing the issues and trends that emerge from an analysis across the regions, drawing from particular countries and cases, while trying to keep their contextual differences in mind. Two insights emerge: first, forest tenure reform has unique characteristics and is distinct from other land reforms; and second, several global dynamics are in some cases driving, and in all cases shaping, these reforms. In this chapter, we discuss the nature of this forest tenure reform, as well as three forces that are shaping the political and institutional context: the recognition of indigenous rights, global biodiversity conservation and democratic decentralization. We note that these forces have come largely from outside the realm of forestry *per se*,[1] particularly in Latin America and Africa, but have nevertheless challenged and influenced it significantly. The chapter concludes by discussing the legacy of these forces on forest tenure changes in light of increasing global concern for the role that forests play in regulating climate.

What is different about forest tenure?

According to the biological definition, forest landscapes – like other common pool resources such as lakes and fishing grounds – hold multiple resources with a wide array of values for various stakeholders, from local dwellers to distant urban folk. Far from providing only timber, forests represent food, shelter and medicine for local consumption. They also hold the genetic biodiversity that may allow us to cure future diseases, house much of the planet's wildlife, regulate freshwater flows across regions and contribute to global climate regulation through their capacity to capture and store the carbon emitted into the atmosphere. They are also sanctuaries for the spiritual renewal of indigenous populations and urban vacationers.

Who has what rights to these resources? Understanding tenure rights as a set of agreements between social actors with respect to a resource (von Benda-Beckman et al, 2006) allows us to break away from the notion of absolute property rights, or a single private owner, and see forest tenure as often involving groups of people with multiple and simultaneous rights and hence a shared interest in a common resource. As with other common pool resources, the task is governing multiple resources in a shared space while maintaining them as renewable resources.

Analysis of the past 20 years of tenure reform in forests requires a departure from premises lingering from agricultural land reforms and assumptions that underlie many administrative approaches to the formalization of property rights. Discussions of property rights and tenure security often overlook the unique set of conditions that distinguish rights to collective or common property resources, like forests, that are particularly important in developing countries. In an extensive review of African cases, Diaw (2005) argues that the layers of embedded tenure systems characteristic of communal or customary lands result in resilience and flexibility that have allowed these systems to survive in the face of persistent antagonism from, and confrontation with, the dominant, more discrete view of land tenure. Recent literature has scrutinized the pitfalls of simplification, noting that overly simplified formal property rights have skewed the underlying social relations that define 'property'. Problems emerge when the logic of previous reforms promoting private individual property is mechanically transferred to tenure reform in the forest or other common pool resources (Cousins and Sjaastad, 2008).

There are several ways to approach the differences. Bruce (1998) lays out a simple distinction in terminology, stating that tenure reform describes legal reforms of tenure rights whether by the state or local communities. It differs from land reform because rather than redistributing land, it more often leaves people holding the same land but with different rights.

According to El-Ghonemy (2003, p34) by the 1960s, 'the accepted sense of the term "land reform" meant the *redistribution* of property and use rights of land for the benefit of landless agricultural workers'. However, he notes that because of the influence of powerful land reform movements in Latin America, 'this previously used narrow English term "land reform" was transformed to agrarian reform corresponding to the Spanish term "reforma agraria"'. The expanded term included transformations in the broader policy frameworks within which land reform might take place and referred to 'the class character of the relations of production and distribution in farming and related enterprises' (Cousins, 2007a, p232). Such reforms were most effective in improving beneficiaries' livelihoods when they fitted into broader policies aimed at reducing poverty and developing productive smallholder agriculture. Targeted credit lines, training and extension programmes, government price subsidies and often a direct role in commercialization all proposed to turn land reforms into the basis for rural 'development'.

Over time, opposition to redistribution mounted and agrarian reform also came to mean government-promoted settlement or resettlement programmes on publicly owned land, land registration, consolidation of fragmented holdings, tenancy improvement and land taxation (El-Ghonemy, 2003). Publicly owned land referred mainly to forests into which peasants and small farmers moved, both spontaneously and through planned 'colonization' or 'transmigration' schemes. These colonists occupied dense forested areas and were offered (or promised) land titles for clearing the forest for agriculture or ranching, in keeping with rural development approaches of the time (Thiesenhusen, 1995; de Janvry, 1981).

Since colonial times in Latin America, land with standing forest cover was officially considered 'idle' land (*tierra ociosa*); in Indonesia, it was considered 'empty' (*tanah kosong*). Throughout the tropics, to gain a rightful claim to either official or customary ownership, the forest had to be cleared.[2] Deforesting the land was seen as a measure of invested labour (e.g. for Cameroon see Diaw, 1997; for Indonesia see Colfer with Dudley, 1993; for Latin America see Clay, 1988), demonstrating the 'social use of land' (and from an official point of view, made it worthy of a title). Underlying this process was the notion that work constituted a basis for ownership, rendered visible through land clearance. This concept of the social use of land has in many areas been supported by law and has become a cultural institution throughout Latin America (Ankersen and Ruppert, 2006), as in many tropical countries.

In Latin America, the forest resource was seen as abundant and the state's objective was to bring forested regions under state control and into productive use. However, colonization programmes often set off new waves of conflict as supposedly empty forests proved to be filled with people unwilling to give up their land and livelihoods without a fight (Schmink and Wood, 1992). Environmental concerns and indigenous territorial rights were largely neglected. In most of Asia and many parts of Africa similar assumptions governed settlement programmes to fill 'empty lands', also already inhabited by local groups, often practicing sustainable, long-rotation agriculture. Though in many places in-migrants were welcomed when populations were sparse, resource-dependent people may become less hospitable as populations grow to densities that endanger the sustainability of local customary systems (e.g. Peluso, 1994).

Also in Asia and Africa, colonial and postcolonial regimes perceived the forest as a scarce resource to be protected from agriculture, poaching and grazing. The creation of forest reserves – along with wildlife reserves for protecting game animals – was the norm and people were excluded and even evicted (Adams, 2004).

In contrast to the agrarian approach described above, the bulk of the current forest tenure reforms have focused on the change in rights that occur *within* forests. These reforms have principally recognized the rights of preexisting forest dwellers, in what are overwhelmingly (officially) state-owned forests. Unlike the case with agrarian land reform, there is an entire array of rights holders bound by a complex web of interests (Meinzen-Dick and Mwangi, 2008), including the state itself. Therefore, forest tenure reform is a far more complex endeavour involving a set of agreements about who has what kind of rights to which forest resources. More often than not, these rights holders' interests vary with the season, climate, price of forest goods or political factors. Forest landscapes, for example, may include rotating agriculture or forest gardens or the seasonal use of forests by nomadic herders. Flexibility in the rules that allocate rights is necessary (Berry, 1993; Barry and Meinzen-Dick, 2008).

Again, in contrast to most agrarian land reforms, where land was titled mainly as individual private property (Alegret, 2003; El-Ghonemy, 2003), forest tenure reform mostly involves granting collective rights but maintaining the state as a principal rights holder. Alienation rights, those that relate to the division and sale of the land, are legally retained by the state. The underlying logic is that forests are ultimately a public good (see Chapter 7), that their subdivision into small units will result in clearing and that only the state can guarantee their permanence (an assumption that has been increasingly called into question).

The implications of those differences for forest tenure reform are several. First, the state remains a rights holder, often playing a central role in forest resource management (see Chapter 3), and the 'enforcer' of exclusion rights. Second, the collectives, communities or groups of communities that receive forest tenure rights are bound in relationships that could be considered co-ownership or co-management of forests with the state. Third, in most countries studied, division and sale of forestland is prohibited, even when titles are granted, and the forests are not legally considered 'property' (are not subject to mortgage or embargo), which lessens the opportunities to use titles as collateral for investment capital or credit. Hence, forestland is not a commodity, in the sense that it does not enter into formal land markets. This constitutes one of the fundamental differences from land reform, where land can be legally divided and sold.[3]

In summary, forest tenure reform involves decisions about access, use, management and exclusion rights – the fundamental rights of forest governance. The struggle for the transfer of real decision-making powers, particularly management and exclusion rights, from the state to communities becomes the centre of our concern. To what degree do communities hold and shape these rights? To what extent does the state undermine or recognize existing decision-making rights and practices? How can the state best organize supporting institutions to enfranchise communities and enhance their benefits from forests? Our analyses of the implications for local community enfranchisement, welfare and forest conditions have identified the following fundamental characteristics of forest tenure reform:

1. land titles or rights are granted with the understanding that the forest resource should be maintained (or restored);
2. tenure rights are essentially for multiple users of various forest resources;
3. in the vast majority of cases, alienation rights to the land are still held by the state;
4. thus the land cannot be legally divided and sold (and is thus not a legal commodity);
5. most of the reformed forestlands are being demarcated and titled as collective or communal properties;
6. this means the recognition of a previously existing collective governance structure and/or the creation of a new one (or a combination).

Context and shaping of forest tenure reform

The current transition in forest tenure began during a period of dramatic political change for many of the forested countries of the south. During the 1980s and 1990s, the Philippines, Indonesia and Brazil witnessed the demise and fall of dictatorial regimes, followed by the construction of more democratic political systems. Guatemala and Nicaragua were embroiled in civil wars and then faced the task of postwar reconciliation and reconstruction. West and Central Africa, specifically Ghana, Cameroon and Burkina Faso, experienced a variety of postcolonial governments from the 1960s to the 1980s, sometimes with substantial upheaval, but generally had returned to constitutional democracies and multiparty politics by the mid-1990s.

During this same period the global development paradigm underwent major modifications as models intended to ensure broad, equitable rural development were abandoned in favour of neo-liberal models of macroeconomic growth (El-Ghonemy, 2003).[4] The earlier rural development models were part of a worldwide undertaking to modernize agricultural production as the engine of rural development, with coordinated financial, institutional and scientific efforts. Land locked up in large and unproductive holdings was to be redistributed to the landless and others, thus providing dynamism in land markets.

By the mid-1980s, structural adjustment policies had forced the contraction of fiscal spending and reversed emphasis on state institutions as the agents of integrated rural development. Programmes that accompanied land tenure reforms waned; the market would become the principal driver of economic growth. Downsizing of the state, fiscal restrictions and monetary policies forced changes in exchange rates and eroded government crop subsidies, leaving rural sectors highly vulnerable to shifts in international trade (see for example Sunderlin and Pokam, 2002; El-Ghonemy, 2003; Thiesenhusen, 1995).

Most countries with significant forestland underwent these macroeconomic changes, which constituted a shift in the development paradigm and the institutions and agencies associated with it. Under market liberalization, agrarian reform was transformed from an act of land redistribution through state intervention to market-based land 'reforms' and land administration projects aimed at formalizing titles and modernizing cadastres and registries (Rosset et al, 2006; Deininger and Binswanger, 2001). For the forest sector, timber trade was liberalized and became export focused, and reduced budgets meant dwindling extension agencies, staff and programmes.[5]

Despite regional differences, the changes in this macroeconomic policy helped set the conditions – particularly for the state – under which forest tenure reform would unfold. During this same period, three additional international trends emerged worldwide:

1 the demand for the recognition of indigenous peoples' rights to their identity and ancestral lands motivated tenure reforms, especially in Latin America;
2 the global drive for biodiversity conservation profoundly influenced how rights and governance in forests have been redefined; and
3 decentralization, in many cases part of structural adjustment programmes themselves, became an important force in forest reform, most evidently in Africa.

Rather than discussing the role and impact of these three dynamics in every region, we present an overview with emphasis on the region in which each force has had particular influence.

Recognition of indigenous rights

Most of the area under forest tenure reform has been in Latin America. As of 2008, roughly 197 million ha had been granted to the continent's indigenous and smallholder communities, all in predominantly forested landscapes (Pacheco et al, 2008c; Sunderlin et al, 2008). Most of this formal transfer of rights has come about through the recognition of indigenous ethnic identity and rights to ancestral lands. Figures on the total indigenous population are inexact, but a 2002 estimate was more than 52 million, or just under 12 per cent of the total regional population (Roldan, 2004).

To put this in perspective, about 120 million people live in rural areas in Latin America (Quijandría et al, 2001). It is estimated that approximately 25 million live in subtropical and tropical landscapes, an important portion of which are covered with forests. Roughly speaking, 12 million people occupy forestlands in Mexico, a large portion of whom are indigenous, 3 million live in the forested landscapes of Central America and about 10 million in the forests of Amazonia, 1 million of whom are indigenous (Kaimowitz, 2003a). Bolivia, Guatemala and Peru have the highest percentage of indigenous population, with 71 per cent, 66 per cent and 47 per cent, respectively, in contrast to Brazil, with the largest land area under indigenous tenure but representing only 0.2 per cent of the population (Roldan, 2004).

The struggle for recognition of indigenous identity and rights dates from the colonial period (which in Latin America had ended by the mid-1800s) and has been a source of strife and conflict, often invisible. In the 1980s, a wave of international recognition of these rights represented the culmination of a sustained global battle – led by Latin America – for the expansion of human rights to consider indigenous claims for ethnic identity and resource rights. At the heart of these legal processes was an attempt to safeguard indigenous cultural reproduction (e.g. language, knowledge, landscape). The rights perspective has shaped tenure reform far beyond Latin America and even beyond indigenous peoples.[6]

Origin and goals

The movements for the recognition of indigenous rights organized over the course of the 1980s and 1990s were first focused on rights of identity. Alliances with international NGOs and legal council brought about achievements in the international sphere, such as the revision of international conventions like the second International Labour Organization (ILO) Indigenous and Tribal Peoples Convention No. 169 of 1989. ILO convention 169 has provisions

> *...requiring ratifying states to identify indigenous lands and guarantee the effective protection of rights of ownership and possession; to safeguard indigenous rights to participate in the management and conservation of resources and to consult with indigenous peoples over mineral or sub-surface resource development.* (Plant and Hvalkof, 2001 pp32–38)

Together with a host of other international legal supports, this convention was eventually ratified by most South American countries with important indigenous populations, giving it force of domestic law in those countries.[7]

Strong national-level advocacy harnessed to the force of these international treaties eventually led to the abandonment of official assimilation policies in most Latin American countries. By the 1990s, constitutional recognition of states as multicultural began to have implications for changes in rights of identity, use of language and political representation and rights to land (Plant and Hvalkof, 2001). These constitutional victories fed the resurgence of indigenous identity throughout the region. But little really changed in practice without secondary legislation, something with juridical 'teeth' that would translate these constitutional aspirations into the concrete manifestation of rights, including rights to land and its resources (Leyva et al, 2008).

Enthusiasm was spurred by the surprise 2001 landmark decision in the Interamerican Court of Human Rights in favour of the Mayangna Indians in Nicaragua, in the Awas Tingni case. When the state failed to consult with indigenous communities over a logging concession on their ancestral lands, the court ruled that the government of Nicaragua had to demarcate and title Awas Tingni's traditional lands. A combination of external and internal pressure, including conditions tied to World Bank loans, led the government to take this step, setting a legal precedent in the country – and a new trend in the region (Larson and Mendoza-Lewis, 2009; Stocks, 2005, see also Chapter 5).

For indigenous peoples, the collective right to land forms an essential part of their identity and is necessary to ensure their cultural reproduction (Bae, 2005). Thus, demands throughout the region have focused on the restoration of the territories that were long inhabited by indigenous and tribal peoples. The concept of *territory*, as opposed to *land*, for indigenous peoples in general refers to the space and resources under their control that enable them to develop and reproduce the social and cultural aspects of their livelihoods. It also reflects the collective aspects of the relationship between indigenous peoples and their lands. This expanded concept of territories was more easily applied to the Amazon and other tropical lowlands, where the contiguous land areas vested in indigenous groups can cover several million hectares, rather than in more densely populated forests. This model also allowed for recognition of an integrated approach to resource management (Plant and Hvalkof, 2001). Although the concept of contiguous territories belonging to groups of people applies in Africa and Asia as well, the scale of such landholdings is much smaller outside Latin America.

Struggles over indigenous lands at the national level often emerged over the right to exclude intruders. The issues included land invasions by colonists and the expansion of forest areas under conservation regimes or extractive industries, such as subsoil mining of hydrocarbons and minerals. Indigenous communities also encountered new restrictions that limited their rights to use their forest resources and in many places met with the negative impacts of extractive activities, such as deforestation related to roads, fires, clear-cutting, water contamination and loss of wildlife and fishing grounds. These are all problems that plague indigenous peoples in Africa and Asia as well, but without the serious policy response.

Latin America's current tenure reforms have taken place almost exclusively in the expansive lowland tropical forests of the Amazon basin and Central America,[8] where there is a very high correlation between indigenous lands and forest cover. These lowlands also include a great variety of ecosystems, ranging from swamps, lakes and river valleys with seasonal flooding to higher-lying savannas and montane rainforest (Chapin et al, 2005; Plant and Hvalkof, 2001). In most cases, reforms have recognized the rights of traditional land users to establish boundaries for claims but disputed their size (often seen as 'too much for too few', Stocks, 2005). The legal process typically dictated that the state establish 'proper land titling systems' for indigenous landownership and recognize customary governance institutions in the territories. The implementation of these mandates, however, has faced numerous hurdles, ranging from the conceptual to the institutional and practical, as well as the disadvantageous position of indigenous peoples in the power structures of their countries.

Implementation and outcomes

The delimitation of large lowland territories in forests previously deemed 'empty' was based on traditional livelihoods, resource uses and customary practices. The holistic and integrated use of the forestland and water system, wildlife and vegetation was accepted as legitimate and legally sanctioned.[9] But this implied the participation of indigenous people themselves in demarcation. Over the next two decades participatory land-use mapping expanded and evolved, accompanied by ethnographic interpretation and geographic referencing based on people's accounts of their land areas and uses (Herlihy and Knapp, 2003; Chapin et al, 2005). International NGOs and eventually support from the World Bank were crucial for developing the tools and increasing indigenous participation in this complex and costly mapping process.

Nevertheless, technical and political barriers hinder the implementation of these reforms. 'Land' reforms in forested landscapes require different approaches from traditional agrarian landownership, yet for the most part the same agricultural institutions and technical agencies remain the principal vehicles for implementation.[10] Usually they rely on the same legal, procedural and even technical norms and mechanisms of agrarian reform (e.g. land regularization, demarcation, elimination of third-party claims, titling and land registration). Even legal conflicts are often channelled to agrarian tribunals, where the rules for arbitration may be inappropriate.

The Bolivian government is in the process of titling nearly 24 million ha to benefit 200,000 indigenous people (Pacheco, 2006). The case of Guarayos exemplifies the processes and outcomes that have ensued in some countries, particularly where indigenous territories are located near agricultural frontiers (see Chapters 3 and 5). Bolivian law recognized community lands and created a formal type of communal property for indigenous people known as a TCO (*tierra comunitaria de origen,* original community land). TCOs were created for individual communities, entire ethnic groups in multiple settlements or even several ethnic groups together. The government carries out a territorial needs assessment taking into account the group's historical occupation of the region, livelihoods characteristics and the potential for population growth. Once the proposed size and shape of the TCO have been presented, the National Institute of Agrarian Reform (INRA) – in theory – prohibits the entrance of third parties establishing new claims.

Existing third-party claims are resolved through *saneamiento* (adjudication), but in practice stakeholders with more economic and political power have been able to influence or slow the process to their benefit. Timber concessions were granted to 11 industries. Ranchers and farmers were beginning to lay claim to expansive territories that indigenous people had treated as commons. Colonist smallholders were encroaching on village space and occupying lands to establish claims. The initial promise of receiving a TCO quickly settled into a long, open-ended administrative process.

INRA failed to take into account the traditional indigenous system of agricultural zones and zones of influence (mostly forestland). It also focused on remote areas with low population that allowed titling to advance more quickly while avoiding zones where land claims were contested – areas of greater interest to local politicians and business. Populous problem areas were not addressed. Most Guarayo families, for example, were left without clear tenure and old and new third-party claimants had time to consolidate their holdings.

Despite such difficulties, the reforms have generally advanced tenure rights for indigenous peoples. For example, in Brazil, indigenous rights were recognized for about 100 million ha involving 500,000 people (Barr et al, 2002) and the state supports their exclusion rights. In Nicaragua, roughly 2 million ha of forestland is in areas being claimed by and demarcated for indigenous territories. The Philippines, too, has begun to recognize indigenous rights through certificates of ancestral domain, such as the one granted to the Ikalahan people in our study (Pulhin et al, 2008; see Chapter 5).

These rights-based reforms have not only benefited indigenous peoples but also opened up opportunities for other claimants, particularly communities whose livelihoods depend on rubber, brazil nuts, *acai*, *chicle* or *xate* and/or whose basis for claims is the traditional, *de facto* possession of forest resources (Cronkleton et al, 2008). The policy responses and mechanisms devised by governments to satisfy the demands of these extractive communities also constitute a central piece of the forest reform. They include, for example, 20 million ha allocated to about 145,000 smallholders and extractivists in Brazil (CNS, 2005). In the Petén, Guatemala, about 500,000ha has been granted through 13 forest concessions to local community groups (Junkin, 2007).

How secure are these new rights? Defending exclusion rights to common pool resources is an inherent problem, and given the relatively small populations and large size of these territories, enforcement is even more difficult. Outcomes depend on the institutional capacity of the residents, the characteristics of the tenure mechanism used and the state's political will to defend the boundaries from incursion.[11]

Additionally, in Latin America (and most countries in Asia and Africa), the multiple interests of the state itself are a central problem. While one state agency grants rights to indigenous groups for extensive forestland, another grants subsoil concessions to industries for resource extraction (hydrocarbons, water and minerals). Rights frequently overlap (Barry and Taylor, 2008), and though legally coherent – the state is usually the owner of all subsoil resources – this duality results in state promotion of incursion into indigenous lands. The same problem occurs in parts of Africa (e.g. Ranjatson, 2009, on Madagascar) and Asia (Moira Moeliono, personal communication, 2009, on Indonesia). Except in rare cases, such as Nicaragua, where communities may have the right to say no to subsoil concessions, depending on interpretation of the law (Larson, 2008), indigenous people have little legal recourse. Yet the incursions stand in stark contradiction to the fundamental nature of the rights they have been legally granted.

Expansion of biodiversity conservation

The second important force shaping forest tenure reforms has been the rise of the global movement for biodiversity conservation. The interests of this international conservation effort stemmed from awareness of the rapid rate of biodiversity loss, particularly due to deforestation in developing countries. It spurred an extensive expansion of forest zoning and conservation areas to protect forests from human intervention by establishing protected areas, biosphere reserves and national parks (Adams, 2004; Sayer et al, 2008).

Origin and goals

From the mid-1980s to the mid-1990s conservation areas grew at a rapid pace in attempts to meet the globally established goal of putting 10 per cent of Earth's terrestrial systems under conservation management. By 2004 the goal had been surpassed, having reached 12 per cent (Bray and Anderson, 2006) and a few world regions went far beyond. In 2003, Central America had almost 28 per cent of its land in protected areas; South America had 22 per cent; Southeast Asia had 16 per cent and eastern and southern Africa had 17 per cent (Chape et al, 2003). The growth of conservation areas, each with its own regime of rights and practices, would serve to restructure the zoning of forest land across many countries. New definitions of what should be set aside, what could support restricted use or what could be logged by whom, and what could be used by local people, redrew the boundaries of the permissible.

The implementation of zoning for conservation occurred more or less at the same time that demands were crystallizing in international law for the recognition of local peoples' property rights (Fisher et al, 2005). The expansion of conservation into Latin America's tropical lowland forests is simultaneous with the indigenous mobilization discussed above, often involving the same lands.[12] Although this was sometimes conflictive, it also spurred attempts at coordination (COICA, 2003).[13] In Africa, conservation, which had originated with the protection of wildlife species for elite colonial game hunting (Adams, 2004), had been transformed over time and by the 1990s had evolved into the protection of large-scale ecosystems encompassing the habitats of all the species of concern.

Roe (2008) describes the late 1990s to the early 2000s, in particular, as a period of 'backlash' against community-based conservation, which had garnered relatively strong support during the previous decade, as both 'conservation and development policy merged around theories of sustainable development'. This backlash marked a return to 'protectionism' based on the arguments 'that community participation is a noble goal but diverts funding away from conservation, and has minimal effect on biodiversity conservation' – a sentiment widely expressed at the 2004 World Conservation Congress in Bangkok.

As a major force reshaping the distribution of rights in forests, conservation policies were often contentious. They would prove to be one of the most highly centralized exercises of defining forest use ever, created not at the national level or at the seat of a colonial power, but from a few cities in a handful of northern, developed countries. The definitions of biodiversity and where it was threatened were determined mostly by northern, and often urban-based, scientists and conservationists. Although such people were at the cutting edge of their fields, their decisions demonstrated a distance from, and ignorance of, the intricacies of social life in and local perceptions of forests.

Biodiversity was defined as 'total diversity and variability of living things and of the systems of which they are a part', including ecosystems, species and genetic diversity (Heywood, 1995, p9).[14] Humans were not included. Underlying premises included the belief that there were large expanses of pristine forests and other landscapes that should be preserved from harm by humankind (Sayer et al, 2008). Few efforts were made to understand the causes of biodiversity loss or the social complexities of these human-modified landscapes; consequently, false assumptions often led to drastic and erroneous measures.

Sayer et al (2008) note that conservationists acknowledged the lack of full understanding of how forest ecosystems really function and thus argued for a precautionary principle, which proved to be a factor in these massive set-asides. Technological advances and the decreasing costs of computer and satellite imagery made it far easier to establish baselines and track changes – such as forest fire and deforestation. Global awareness of macro-level forest dynamics jumped and was soon linked to understanding the role that forests (and other landscapes) played in maintaining a balance in global climate.

Much of this growing knowledge was housed in a few large, mostly US-based NGOs whose staff became the organizers and promoters of this new conservation agenda. During the 1990s these organizations witnessed enormous growth and became increasingly powerful in influencing the policies of national-level agencies and institutions in charge of forest management around the world (Khare and Bray, 2004). Together, these international conservation agencies came to enjoy budgets often equal to or larger than their counterpart environmental ministries in many developing countries, themselves suffering from fiscal downsizing. Between 1998 and 2002, the combined revenue of the three principal NGOs – Conservation International, The Nature Conservancy and World Wildlife Fund – made them the second most important conservation actors, after the multilateral banks' conservation programmes (Khare and Bray, 2004).

Implementation and outcomes

After designating 'hotspots' or eco-regions, the conservation agencies next sought to gain the assent of national governments for the adoption of conservation areas, with a promise of financial and technical support. Some areas became only 'paper parks'. In others, large-scale, wholesale rezoning of forest access and use created conflicts at many levels, especially with indigenous peoples (Colchester, 2000b, 2004).

In Central Africa the advance of the conservation regime met with little visible resistance as governments in some cases adopted conservation criteria into their forest zoning, either ignoring or evicting local forest peoples (Brockington and Igoe, 2006; Cernea, 1997, 2006). In Southeast Asia, protected area policies were often based on coercion and criminalization of people living near forests, and sometimes their eviction as well (Inoue and Isozaki, 2003); such a process is currently underway in northern Laos (Fitriana, 2008). Some Mexican and Central American conservation efforts redesigned whole rural landscapes: the large ecosystem approach to conservation drew boundaries for set-asides or limited resource use, sometimes including small towns and whole villages within their perimeters (Secretariat of Environment, Natural Resources and Fishing – SEMARNAP, personal communication). The margins of error were gargantuan, and legal battles and social upheaval often ensued (Plant and Hvalkof, 2001; Colchester, 2004).

Particularly egregious policies, such as the eviction of local peoples from protected areas – now disputed (Maisals et al, 2007) – have been more common in Africa and Asia than in Latin America (Brockington et al, 2006; Cernea, 1997, 2006; Adamson, 2003, in Bray et al, 2005). Dowie (2005), referring to 'conservation refugees', documents forced removals on three continents and discusses the growing discord between communities and conservation organizations, which are increasingly seen as the new colonizers. Displacement of peoples for conservation purposes is still underway in many parts of the world (see for example Ghate and Beasley, 2007; Baird and Shoemaker, 2005), though as Curran et al (in press) argue, some authors have exaggerated its prevalence.

In other contexts, where resident communities were 'found' and their claims to natural resources prevailed, conservationists promoted the imposition of management regulations (Bray and Anderson, 2006). By 2000, some conservationists realized that local forest dwellers ware far more prevalent and numerous than originally estimated. Oviedo (2002) reports that about 86 per cent of national parks in South America are inhabited, mostly by indigenous and traditional peoples. Globally, more than 1 billion people (at least 25 per cent of whom are malnourished) live in global biodiversity hotspots, the 25 large-scale biodiversity priority areas identified by Conservation International, subsisting on less than US$1 per day (McNeely, 1999, in Molnar et al, 2004).

Regulations on people living in these areas include prohibitions on the hunting of certain wildlife species and the suppression of other basic practices of forest–agricultural systems. For example, from pre-colonial times until the present, swidden agriculture has been a central land-use strategy by peoples living in tropical forests around the globe. Groups of people (e.g. communities, tribes) typically recognize a territory that belongs to them (what Diaw, 2005, has called 'collective property') and forest clearing, planting and regeneration are both livelihood practices and a way of staking their claim to a part of the forest (see Chapter 4). Highly criticized by conservationists as destructive, shifting cultivation may have a role in creating and maintaining resilient, adaptive livelihoods and forest systems (Kerkhoff and Erni, 2005; Cramb et al, 2009).

Finding the forests more populated than expected, facing the demands of indigenous and other customary land claims, both on the ground and in the courts, and calculating the real costs of these exclusion schemes have led conservation NGOs and their national counterparts to move towards co-management approaches that incorporate greater participation of local communities in forest management. In some cases, these populations allied with conservation organizations, and the environmental agencies with whom they worked, against their common 'adversaries' (loggers, miners or ranchers). The establishment of the Maya Biosphere Reserve, straddling Guatemala, Mexico and Belize, evolved from outright hostility and the threat of social unrest, through clashes among NGOs, the government environmental agency and local forest dwellers (in the Guatemala portion), to a common agreement. Communities organized to protect their rights to forestland and resources and pushed for replacing the industrial timber concessions with community concessions (Monterroso and Barry, 2007; see Chapters 3 and 6). The result is a successful co-management arrangement between conservationists and community foresters (Nittler and Tschinkel, 2005).

Though limited mostly to temperate forests in developed countries, an important contribution of conservation efforts to forest management has been the emphasis on expanding the definitions and application of sustainable forest management criteria. One stream has focused on reduced-impact logging (de Camino, 2000; CIFOR, 1999); another has explicitly linked community management, sustainable forest management and conservation concerns (see for example Ritchie et al, 2000; Colfer and Byron, 2001). The establishment of principles for forest certification that recognized the social and economic realms helped raise awareness within the forestry community of the needs and possible contribution of local forest peoples (Molnar, 2003) and the potential of building on local governance systems (Gibson et al, 2000).

The fundamental issues of addressing global interests in biodiversity conservation and their impact on forests and forest-dwelling peoples remain only partially resolved, however. Roe (2008) summarizes the concerns as:

1 the impact and accountability of the activities of the big international conservation NGOs;
2 the apparently increasingly protectionist focus of conservation policy and the implications for communities in and around protected areas; and

3 the current lack of attention to biodiversity conservation on the development agenda, with the prioritization of poverty reduction and carbon sequestration.

Sayer et al (2008, p3) point out that 'biodiversity presents special challenges in determining optimum arrangements for use and ownership of forests. Biodiversity has certain values that accrue primarily to the global community while local owners and users of the forest lack effective mechanisms to profit from these values'. With the resurgence of interest in forests under new climate change mechanisms, lessons from the experience of biodiversity conservation and its implications for forest tenure reform and governance would be well heeded.

Decentralization and natural resource management

The third factor shaping tenure reforms is democratic decentralization, defined as the transfer of power and resources from the central government 'to authorities representative of and accountable to local populations' (Ribot, 2004, p9). It is associated with the development and strengthening of local elected governments but often blends other arrangements. Decentralization as a global policy trend has been promoted in the name of local democracy by international organizations such as the World Bank, particularly since the late 1980s. Not all decentralizations have implications for natural resource management or for tenure rights; this section focuses on those that do.[15]

Decentralization has not been the primary driver of forest reform in Latin America. In Bolivia, however, both departmental and local governments participate in forest administration, and local governments can establish municipal reserves in up to 20 per cent of public forests to give as concessions to local logging associations. In Asia, decentralization has been central to forest reforms in Indonesia, where important powers over forests were granted to local governments, some of which have since been rescinded (Resosudarmo, 2005; see Chapter 4); the impacts on communities' rights have been minimal, however. Community forestry in Nepal began with decentralization to *panchayats* (local governments) but has since shifted to devolution to community user groups (Agrawal and Ostrom, 2001). In the Philippines, community forestry emerged out of a wave of democratization policies after the fall of Ferdinand Marcos, but decentralization as a specific policy came somewhat later, granting a role to local government in policies that were already underway (see Pulhin et al, 2008; Magno, 2001). This section focuses on Africa, where the reforms are characterized by the nature of the interface between statutory and customary regimes.

Origin and goals

Decentralization is not a new phenomenon (Kuechli and Blaser, 2005; Sasu, 2005). What is new about current decentralizations is the emphasis on 'democratic decentralization', particularly through the formation of autonomous local governments (in its 'ideal form', Ribot, 2002), as well as a discourse promoting participation in decision-making, participatory democracy, pluralism and rights (Conyers, 1983). The theoretical benefits are well known by now: decision-making closer to local people should be more equitable, efficient, participatory and accountable and, possibly, ecologically sustainable. In practice, however, decentralizations have often been implemented in response to economic or political crises or pressure from donors and for reasons not always related to these goals. Ribot (2004, p8) summarizes the main objectives as government downsizing or, alternatively, consolidation ('shedding risks and burdens'), promoting national unity, improving service delivery, increasing local participation and democracy and strengthening local government.

Agrawal and Ostrom (2001, p492) argue that 'decentralization can be said to have occurred only when governments devolve property rights over resources' in such a way that some level of decision-making over management, exclusion and alienation (the 'collective-choice level of analysis') is granted in the local arena. At the same time, there must be a set of 'constitutional-level' rules that secure local people's right to make these decisions. 'Simply granting rights to undertake operational-level actions is insufficient to justify claims of decentralization' (Agrawal and Ostrom, 2001, p492). Nevertheless, in practice forest decentralization policies sometimes result less in devolving local control and more in maintaining or even increasing state control over local forests and forest communities (Becker, 2001; Sarin et al, 2003; Elías and Wittman, 2005; Schroeder, 1999).

Policies implemented in the name of democratic decentralization almost by definition affect tenure rights in public forests by altering the distribution of decision-making powers in the local arena; this is particularly true where customary tenure rights are widespread. In Africa, the state formally owns virtually all forestland in a number of countries (RRI, 2009), but some 60 per cent of the total forest estate is 'off-reserve' or not formally 'classified' by the state; in these areas, 'customary and other unregistered forms of tenure dominate' (Alden Wily, 2004). Decentralization policies in Africa present a mixture of maintaining or increasing state control (e.g. Mongbo, 2008), usurpation of power and benefits by customary authorities (e.g. Ntsebeza, 2005) or elites (e.g. Oyono, 2005b) and, at times, devolving rights to local communities, particularly in unclassified forests (Alden Wily, 2004).

Implementation and outcomes

Under colonialism, policies of 'indirect rule' for anglophone or 'association' for francophone Africa in a sense set up parallel societies, described by Mamdani (1996) as 'decentralized despotism', whereby Africans lived under customary law and authorities and Europeans and urban citizens lived under statutory civil law (Ribot, 2002). Though lauded at the time by some as a way to promote self-determination (Mair, cited in Ribot, 2002), indirect rule and similar colonial policies were later repudiated by many as racist, cruel and unjust.

The system was implemented by incorporating chiefs or other 'customary authorities' into the administrative structures of the colonial state. These authorities, which sometimes were not, in fact, customary at all, permitted the state's administrative control and management of rural affairs. Chiefs were empowered by colonial authorities to allocate customary land for local use, which became the basis for their power (Mamdani, 1996).[16] The legitimacy of both local governments and customary authorities was undermined by 'the coercive abuses of the colonial state' (Ribot, 2002).

Colonial forest policies were based on state control of forests and the incorporation of scientific forestry principles (see Chapter 7). French policies in west Africa, for example, established state ownership of forests under two classifications: classified forests, under direct control of the state, and protected forests, all others that were not privately owned; communities had the right to subsistence use of forests, but the state managed all commercial use (Becker, 2001). Chiefs were in charge of local land allocation but had no rights to manage forests (Ribot, personal communication).

In the first two decades following independence the top priority of many new postcolonial governments was to consolidate central control over the country. In many cases, this was a time of great political turmoil. Decentralizations that occurred during this period were 'without exception' in the form of de-concentration[17] (Ribot, 2002). It was in the 1980s and 1990s that the discourse of democratic decentralization took hold. Today, the commitment to greater 'devolved governance of society and its resources' has been written into '20 or more new National Constitutions across the continent'; also, since 1990, 41 out of 56 African states have drafted new forest laws (Alden Wily, 2004, p2).

The implementation of this forest sector decentralization has taken several paths in shaping local tenure rights. Here we focus on general trends and the most important issues raised by our case studies in Cameroon, Ghana and Burkina Faso. In general, reforms overall have continued to maintain centralized decision-making power over forests. This is best summarized by Ribot (1999, p23):

> *The current decentralization and participatory movement is devolving state-backed powers that are still administratively driven and locally administered by quasi-local quasi-representative bodies...In the context of ongoing administrative management of rural areas, participatory projects and laws create privileges to be allocated mostly by foresters and councilors, often with burdensome responsibilities, rather than rights for communities and individuals that the state would defend. Such projects and laws administer local programs rather than devolve control. They back centrally chosen and/or non-representative powers rather than supporting representative systems of local governance.*

Rights to classified or priority state forests, which tend to be the richer and more valuable, have been granted to communities at best through co-management arrangements in which the state plays the central role in all management decisions. Communities are somewhat more likely to gain more substantial decision-making powers in less valuable forests, where communities have enjoyed greater *de facto* or customary control previously; in a few countries these have even begun to be registered as common property to communities (Alden Wily, 2004). In other cases, however, forest administrations have only granted local rights while also increasing their own management role (e.g. Oyono et al, 2006).

With regard to tenure rights, the central issue of concern is the extent to which reforms increase or reinforce rural people's rights to land and forest resources. On the one hand, however, 'recognizing customary rights' sometimes involves reinforcing the power of customary authorities who may fail to act for the benefit of communities; on the other hand, granting greater power and oversight to elected local governments may undermine customary rights and practices through greater state interference. Based on his research in Mali, Benjamin (2008, p2260) found that the superimposition of modern legal institutions on community institutions through decentralization created ambiguities that 'can undermine both the authority of nascent local governments and the performance of customary institutions'.

In our cases, the types of reforms studied involve benefit-sharing arrangements, community forests and community concessions. In Ghana, the constitution provides for a portion of revenues from logging to be returned to the local sphere, through both local governments and chiefs. The larger portion is provided to chiefs, without clarity regarding their obligations to spend these funds to benefit communities; most use them for personal gain (Marfo, 2009; Chapter 5). Cameroon's community forests are granted not to customary authorities but rather to newly created management committees; these are often usurped by local (and sometimes external) elites, who can finance and manoeuver the complicated requirements for approval (Oyono et al, 2008).

Concessions in Burkina Faso vary based on the forest classification, with greater room for customary practices and local decision-making in those forests of less interest to the forestry administration. These unclassified forests are under elected village councils whose decisions are subject to approval by the local government (Kante, 2008; Chapter 5). It remains to be seen whether this approach will succeed in respecting and reinforcing, rather than undermining, customary rights and practices, but at least some people trust their traditional authorities more than local governments (see Diaw, 2009).

One of the paramount issues is how to respect people's customary rights and practices while not reinforcing unaccountable customary authorities who may usurp benefits intended for communities. As Alden Wily (2008, p46) writes:

> *'Tradition' (or custom) especially need[s] to be put in context, for it is not necessarily the substance of old rules or even the identity of rule-makers that needs embedding in statute but that such arrangements derive from the 'communal reference' – the fact that*

> *local community, not state is the source of decision making, norm making, regulation and enforcement.*

The question is not which entity is a better representative of local people but whether it is possible to negotiate fair, workable solutions grounded in such a 'communal reference' (e.g. Benjamin, 2008).

Discussion and conclusions

Fundamentally, in today's reforms, rights are granted to a collective rather than to individuals, alienation rights over the land are not granted, the state maintains an important ongoing role in forest management and the forest is expected to remain intact. Reform is aimed at three objectives simultaneously: addressing demands for greater rights from communities already living in forests, improving livelihoods and promoting conservation.

Three global forces or dynamics have shaped these reforms: demands for indigenous rights, conservation and decentralization. Indigenous rights have played a central role in driving reforms in Latin America; decentralization has been the principal driver in Africa. Conservation has played a role globally and all three dynamics have been important to some degree in Asia. Each of these forces, in each national (and local) context, shape the playing field by influencing who has which powers and rights over which resources.

The *indigenous rights struggle* brought the criterion of rights into tenure reforms globally, even if the initial intent involved ethnic identity, ancestral occupation and use of forestlands. In practice, this opened the way for recognition of non-indigenous forest-dwelling peoples' rights as well. The rights of forest dwellers became the starting point for determining the location and extent of forest access and use areas, and local people were recognized as agents of conservation. Some indigenous groups have won a certain degree of recognition of self-governance, as well as respect for their culture and identity. Historical land-use practices were used to determine the boundaries of territories and sometimes these were drawn through participatory mapping practices. Indigenous perspectives also introduced greater recognition of multiple uses and more holistic views of forests.

Some challenges have not been addressed, however. For example, land-use mapping can define the perimeters for demarcation and titling but has not had the capacity to recognize – for the use of communities themselves – internal and socially embedded systems of customary rights to land and resource use. Also, territorial demarcation has often raised issues regarding representation and decision-making on behalf of the collective, since it has almost always created a demand for new levels of governance, as well as difficulties in defending the borders of the territory. These rights reforms have also been introduced in part to respond to a moral commitment. They fail to offer incentives for effective forest resource management and market participation, where these are desired by communities. Finally, the state undermines these commitments by giving out subsoil rights overlapping indigenous land and resources and fails to honour or back exclusion rights.

Forest conservation efforts have ensured that a broad set of forest values are respected by reforms through the emphasis on biodiversity and the global value of forests. It has introduced and promoted the application of broader scientific criteria into the framework for forest management through the development of new principles for low-impact logging, certification schemes and formal community management. At times, the alliance of conservation interests and local groups demanding forest rights has been critical in winning state and donor support for reforms.

The recent global approach to conservation, however, has tended to stem from elite and external determination of forest resource rules, with little understanding of the needs and rights of forest dwellers, their historic role (at times) in conservation and conflicting definitions of biodiversity. Top-down zoning and regulation by distant policy-makers through rigid categorization promotes exclusion, restricts and undermines customary rights and management practices and overburdens livelihood options for both subsistence and market access. Protection without people has not worked and attempts to create such formal management regimes have sometimes destroyed local governance and management capacity without providing an effective alternative.

Decentralization policies have also had mixed results. Though not designed as an instrument for tenure change, depending on its goals and how it is implemented, decentralization affects the realm of local rights over forest resources. In some countries, decentralization has contributed to greater understanding and recognition of customary rights and practices of local resource management. It has the potential to promote greater and more democratic local decision-making as well as to address complex and overlapping tenure regimes.

Yet decentralization has often perpetuated or even deepened and extended a colonial-type state role in local forest management. It has also encountered significant challenges at the interface of statutory and customary practices and rights, sometimes imposing the former or overlaying it on to the latter, failing to protect rights and promoting greater insecurity. Similarly, decentralization has brought to light complex issues of customary authority: some customary authorities are more legitimate or contribute to better forest management practices, but others are autocratic and unaccountable and usurp benefits and decision-making intended for communities. So far there has also been little acknowledgement or successful integration of indigenous knowledge and management practices with other management efforts. Reinforcement of only upwards accountability in the decentralized structure works against local development and fosters corruption. Decisions over tenure rights are often highly vulnerable to policy change, thus promoting insecurity.

In this examination of the forces that shape today's transition in who has what rights over forest resources, we see the absence of a shared and full understanding of the nature of and challenges implied in the reform. The forest as a social and ecological construct with multiple values of local, national and global importance is abandoned, maybe has never been grasped. Each 'uncle', whether the indigenous rights movement, decentralization processes or conservation agents, brings its concerns and goals to the fore and attempts to 'manage' and benefit from forests from its vantage point. Yet the forest, in its multiplicity of functions and uses, and forest dwellers remain only partially tended to, often undermined and certainly far from being empowered to play their potential role. With the advent of new global interest in forests stemming from concerns over the role they play in climate change, the question is whether this new force will be capable of integrating a global understanding of what a forest and its peoples mean for their preservation, and how getting the rights right will make a difference in maintaining both – or whether it will be another, distant uncle.

Notes

1. By 'forestry' we mean forest science, forestry agencies and policies designed specifically for the forest sector.
2. This is not to be confused with swidden agricultural systems where land clearing formed part of a forest, though this kind of clearing also established claims within the local or customary systems.
3. Under the current neo-liberal view, land is seen as a commodity rather than a social institution (El-Ghonemy, 2003).
4. In general, the antipoverty programmes developed under neo-liberal reforms were designed in response to the negative and often dramatic impact of structural adjustment policies. Conceived of as 'safety nets', funds were targeted at the most vulnerable populations. Design and implementation varied by region, but these funds were expected not to propel rural areas into modernized development, but rather to guarantee a flow of welfare supports.
5. Effects varied enormously in Asia, Africa and Latin America in terms of the size, budget and political power of the forestry institutions and agencies.
6. See Colchester (2000a, 2000b), and Campese et al (2009) for a recent collection on rights-based approaches in conservation globally. Swiderska et al (2009) summarize pertinent international legislation.
7. As of 2004, 17 countries had signed: Argentina, Bolivia, Brazil, Colombia, Costa Rica, Denmark, Dominican Republic, Ecuador, Fiji, Guatemala, Honduras, Mexico, The Netherlands, Norway, Paraguay, Peru, Venezuela (GTZ, 2004).
8. Mexico is the exception; the Revolution of 1917 and subsequent 40 years steadily gave recognition for indigenous communal lands and some customary forms of governance (Bray et al, 2005).
9. In theory, this would include the use of swidden systems within these large territories.
10. This is mirrored in the multilateral funding agencies, where little coordination exists between Land Administration and Forest sectors.
11. For example, in Brazil the National Foundation for Indians (FUNAI) plays a very active role in maintaining the outer boundaries of indigenous territories, and the Brazilian Environmental Agency (IBAMA) plays the same role in the case of extractive reserves. In contrast, in Bolivia indigenous people have no similar state agency looking after their interests.

12. For a visual reference, see the first digitized map of the officially recognized indigenous lands and conservation areas and their overlap in the Amazon basin: www.raisg.socioambiental.org (last accessed September 2009).
13. A well-known example of this was the coordination and support given by the conservation organizations to Chico Mendes and the rubber tapper movement in Brazil, which mobilized global awareness of the fate of the rainforests.
14. Redford et al (2003) categorized the sometimes quite different definitions of biodiversity implied in different approaches. Biodiversity targets and objectives can range from valuation of the presence of a species, ecosystems and ecological processes, scenery and landscape integrity to biodiversity measured as an intrinsic good or something of current or future utilitarian value.
15. Over the past decade researchers have analysed the impact on forest governance of this recent wave of decentralization (e.g. Colfer and Capistrano, 2005; Colfer et al, 2008a; Ferroukhi, 2004; German et al, 2009; Ribot and Larson, 2005; Larson and Soto, 2008).
16. This is similar to what Latin America underwent during the Spanish colonial period, a phenomenon referred to as *caciquismo,* with indigenous leaders co-opted to serve as a liaison class between the communities and the colonial powers.
17. The transfer of powers to branch offices of central government entities.

Part II
The Transfer of Tenure Rights

3

The Devolution of Management Rights and the Co-Management of Community Forests

Peter Cronkleton, Deborah Barry, Juan M. Pulhin and Sushil Saigal

The granting of greater local control over forests through tenure reform has created more opportunities for community forestry and possibilities to improve local livelihoods. However, these reforms entail complex shifts in multiple rights and responsibility as well as changes in the relationships between formal rights recipients, government agencies and other stakeholders. How rights are defined and the institutional arrangements that assign and regulate them can either facilitate the adoption of forest management or discourage local participation in forestry. To increase understanding of how tenure reform has shaped forest use by communities, this chapter will disaggregate the bundles of rights and tenure systems in local contexts to explore how these factors influence community forestry models. It will use four cases of forest tenure reform – in Guatemala, India, the Philippines and Bolivia – to illustrate the complex dynamics and multifaceted nature of these changes.

When transferring forest tenure rights to rural communities, states rarely relinquish full control over resource management. Normally negotiation and struggle define the community forestry institutions that emerge, determine how benefits are distributed and, ultimately, how well forests are maintained. Usually, the state retains ownership and control over forest resources, either authorizing use of public lands or requiring forest users on private or communal property to operate under government supervision and within the normative frameworks it defines. Often, these arrangements recognize some existing resource use embedded in local livelihoods and customary practice, but they also introduce new governance standards attempting to restrict certain management decisions and behaviours. Because management rights are only partially granted and government reserves its role in decision-making, the resulting models require collaboration between communities and state agencies. Situations in which authority, control, responsibilities and benefits are shared to varying degrees are known as collaborative or co-management systems.

From the government's point of view, there are several reasons to adopt co-management approaches rather than relying exclusively on command-and-control strategies. Co-management continues the state's ability to structure and regulate the forestry sector to ensure that both conservation and development agendas are addressed. More importantly, co-management can engage forest-dependent people and bring local knowledge of social and environmental conditions into decision-making. Theoretically, by creating opportunities for direct local benefits from sustainable resource use, such an approach can provide greater motivation to maintain forests. Recognizing local roles through co-management can also alleviate tension with local constituents who had been adversely affected by top-down command-and-control enforcement of environmental law.

The analysis in this chapter emphasizes the nature of management rights extended to communities and examines two central questions: who receives the management rights and how are they defined? How are responsibilities and decision-making powers distributed within co-management systems? We focus primarily on those rights related to resource management and illustrate how powers and responsibilities are balanced between stakeholders and how these arrangements have emerged. Normally, rights were partially devolved to community groups but important controls were retained by government agencies. The resulting co-management systems have produced both positive and negative outcomes. On one hand, the systems involve local stakeholder groups in management, advance forest conservation goals and generate opportunities for forest benefit sharing with local people. On the other, regulatory frameworks and requirements used by the government to maintain control exclude potential participants and cause many community groups to depend on subsidies to join the systems.

Background

Forest tenure reform takes place in landscapes that are often composed of multiple stakeholders, competing interest groups and distinct public agencies holding rights and claiming control over land and forest resources. The transfer of tenure rights is further complicated because it is multifaceted, involving different types of tenure systems and rights bundles. To examine these changes, it is necessary to distinguish the holders of rights, normally conceptualized within one of three tenure systems: public property, communal property and private property. These systems commonly overlap (Feeny et al, 1990) and communities are often mosaics of public, common and privately held land. Frequently, the tenure changes depend on the resource that is being considered; for example, in numerous countries, forests and subsoil resources are held by the state as public property regardless of whether they are located within communal or private properties.

Tenure rights should be conceived of as a 'bundle of rights' (Schlager and Ostrom, 1992) consisting of access, withdrawal, management, exclusion and alienation rights. These rights grant powers of choice and action to the rights holder. All of these rights are defined by rules. The final three rights (management, exclusion and alienation) are considered 'collective-choice rights' or decision-making rights, since they allow the rights holder to define rules and standards for exercising other rights, such as who has access to the resource or how the harvest of a resource takes place (Schlager and Ostrom, 1992). Because they allow the rights holder to establish new rules or adjust those that exist, these rights are crucial for allowing resource users to adapt to changing conditions affecting resources or their livelihoods.

The nature of management rights granted is crucial for analysing how community forest management functions. Resource management should be understood as a collection of decisions, practices and concepts that involve decision-making beyond the immediate resource use. In a forestry context, this could include decisions to manipulate a resource, such as investing in silvicultural practices to encourage regeneration or to avoid damaging future crop trees, or planning or organizing future activities. Although management and withdrawal rights are similar, the primary difference is in the level of decision-making power. A holder of withdrawal rights can harvest resources but only within defined parameters, with no power to decide how, when or what resource use will take place in the future and little control over others who share withdrawal rights. With management rights, the rights holder can make such decisions. Management rights are closely tied to exclusion rights. Taking advantage of management rights entails investments for future use of a resource, but for these investments to be worthwhile, the stakeholder needs to have the authority and ability to exclude outsiders and others who would not comply with management rules.

In practice, transfer of the entire bundle of rights rarely occurs. Instead, some rights are withheld by the state or are not offered without official oversight or control. In fact, these rights are often not held in their entirety by any one individual, are frequently shared among groups and/or different rights can be held by distinct individuals. According to Meinzen-Dick and Mwangi (2008), property often entails 'webs of interest' that combine public, collective and individual rights over resources. An example would be a collective system where rights may be held by a communal authority that determines the allocation of access rights, but resource use takes place at the individual or household level rather than communally. Even where collective and individual property is recognized, these often involve state claims of authority, particularly in relation to subsoil or forest resources. A common trend in tropical forests is that the state retains or restricts alienation rights while recognizing other rights for indigenous or traditional forest peoples.

Depicting the property arrangements as a matrix with rights holders on one axis and bundles of rights on the other can produce a schematic 'tenure

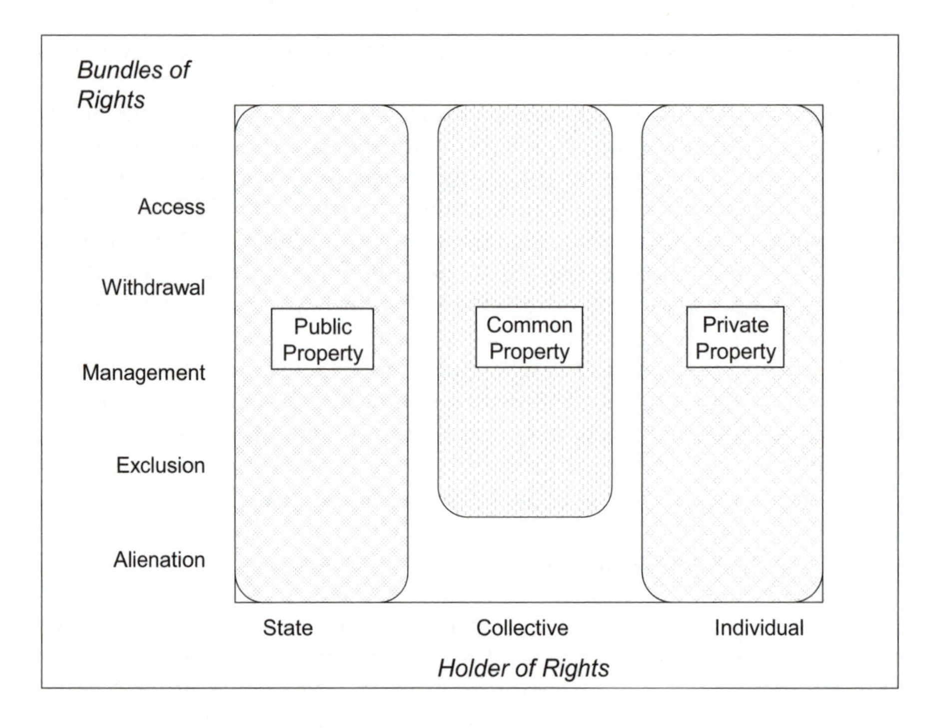

Source: Meinzen-Dick (2006)

Figure 3.1 *Classic property rights systems*

box' (Meinzen-Dick, 2006) that captures the multifaceted characteristics of a property rights system (see Figure 3.1). Disaggregating rights and rights holder configurations allows for analysis of how systems change with devolution or other catalysts. It provides a useful tool for diagramming how rights and rights holders are situated within community forestry systems.

Under forest tenure reform, the management rights granted by the state usually represent a partial devolution of some decision-making powers. When dealing with forest resources, the state typically maintains control through oversight or places restrictions on how community-level actors use and benefit from forest resources. For example, a community given the right to commercially manage timber may be able to choose what portion of its forest to manage, what trees to harvest and how to carry out the harvest; however, its decisions must be approved and comply with management norms established by the state. Although it could seem counterintuitive to devolve rights and then limit them within regulatory frameworks, such give-and-take is part of a negotiated struggle between conservation and development proponents. It has resulted from the realization that local people have roles to play in forest conservation and development. Past attempts to exclude forest-dependent populations from forests have proven ineffective and at times counterproductive. In fact, efforts based solely on command-and-control schemes can limit the adaptation that is the key to resiliency and could undercut conservation goals (Armitage et al, 2009). In addition, transferring rights to community-level stakeholders provides access to detailed local knowledge necessary for good management decisions and involves local interest groups that could do a better job than forest bureaucrats making standardized decisions in distant offices.

The transfer of management rights under forest tenure reform usually produces community forestry models involving co-management arrangements. The widely used concept of co-management is generally agreed to be 'the sharing of power and responsibility between the government and local resource users' (Berkes et al, 1991, cited in Carlsson and Berkes, 2005; Fischer, 1995). Rather than a static state, co-management should be understood 'as a process in which the parties and their relative influence, positions and activities are continuously re-adjusted' (Carlsson and Berkes, 2005, p67). Finding the right balance in sharing rights and responsibilities can be a struggle. As a process, co-management consists of negotiation, bargaining or mediation and provides a venue for problem solving and learning. Ideally, it combines the strengths and mitigates the weaknesses of each of the partners involved (Singleton, 1998). However, co-management can also generate unintended outcomes for communities. For example, onerous restrictions on resource use can discourage participation in the system, stifle innovation or even exclude some stakeholders. Also, because the state is not a single entity, attempts at collaboration can be undermined when different government branches or agencies have contradictory policies.

What does co-management mean for community forestry? Usually the state maintains ownership and nominal control over forests (i.e. full retention of alienation rights and a partial role in others). Rights for management operations are granted conditionally, requiring compliance with regulations. Within the state's parameters, the rights holders are allowed decision-making power. In general, the types of decisions that need to be made include decisions about:

1. who participates in management and how they participate;
2. which resources will be managed and how; and
3. who benefits from management and how.

Normally, legislation requires community groups to form management organizations (to take responsibility for the resource and the impacts of management practices), develop formal management plans and comply with technical standards. The sharing of decision-making does not eliminate command-and-control aspects, since the state typically requires approval, carries out field inspections and imposes restrictions on transport and sanctions for noncompliance. As a result, the institutional structures set up to allocate and control management rights can be very complex and entail high transaction costs for both communities and governments. Variation in co-management systems is strongly influenced by the decision powers that are granted or retained by the state and which stakeholder bears the burden of transaction costs or management risks.

Case studies

This section will examine forest tenure reform and its impacts on four community-level stakeholders: community forest concessions in the Guatamalan Petén, timber grower cooperative societies in northern India, a forest resource development cooperative in the southern Philippines and indigenous forest management associations in lowland Bolivia. To analyse how co-management arrangements in community forestry systems are organized and function, this chapter uses the tenure box to illustrate how reforms shifted rights bundles. A second diagram describes how the framework determines the set of stakeholders holding management rights and in turn establishes relationships and parameters defining co-management systems.

Guatemala: Community forest concessions

The Guatemalan Petén was the site of the transfer of tenure rights to community-level organizations in the buffer zone around a major conservation area. The region has experienced a dramatic shift in forest property rights, where forest-dependent communities went from holders of weak customary rights to holders of forest concessions credited with improving regional governance. Improvements in local livelihoods and forest conservation have resulted, but the co-management system still faces challenges. Community groups have had to respond to contradictory strategies of state agencies with jurisdiction over the region and at the same time defend the community concession model from proposals that threaten to take back the rights gained in the reforms.

For many years, the Petén was one of Guatemala's most geographically and politically isolated regions. Throughout the 20th century, consecutive waves of official and spontaneous settlements brought an ethnically and socio-economically diverse population that today has reached 367,000 inhabitants, mostly migrants from other departments. Initially, extractivist communities were established in the 1920s to harvest valuable non-timber forest products (NTFPs) such as *chicle* gum (*Manilkara* spp.) and the ornamental palm *xate* (*Chamaerdorea elegans, C. oblongata* and *C. ernesti-augustii)*. In the 1960s, state-sponsored colonization policies brought indigenous and ethnically mixed settlers to the region, most of whom practiced swidden agriculture and ranching. Other stakeholders moved into the Petén in search of petroleum, precious minerals and timber. These new arrivals greatly accelerated forest conversion.

Historically, the government has had a weak presence in the Petén, and its policies were often ambiguous or contradictory. Although peasant communities used the forests, the government also granted rights to others to harvest NTFPs and timber. In 1989, the creation of the National Commission for Protected

Areas (CONAP) signalled the movement of an environmental agenda to the forefront of state policy in the Petén. In 1990, much of the northern Petén, which had been treated as a forest reserve, was converted into the Maya Biosphere Reserve.

The original territorial scheme encompassed 2.1 million ha with a nucleus under strict conservation rules, a large multiple-use zone where sustainable timber and non-timber harvest was allowed and a buffer zone to relieve pressure on the reserve. CONAP's mandate was to halt illegal logging, stop forest conversion for agriculture and ranching, stop the sacking of archaeological sites and end illegal traffic in drugs, fauna and migrant workers (Nittler and Tschinkel, 2005). The initial restrictive policies in the reserve for land and resource use triggered serious conflicts with the local population. As competing interest groups resisted, it became more difficult for the government to enforce conservation policies that entailed the exclusion of important forest stakeholders (Monterroso and Barry, 2008).

The main source of local opposition was an organization founded in 1995 by community groups to pursue forest management rights that became known as the Association of Forest Communities of Petén (ACOFOP). In response to this resistance, government policy shifted to an approach that delegated management rights to some forest-dependent stakeholders: 25-year forest management concessions were granted to six local communities within the multiple-use zone, six communities bordering it and two local forest industries. The establishment of community concessions created the collectivization of forest rights and established a common property system over 426,000ha of forestland (Monterroso and Barry, 2008). International conservation groups funded heavily by The United States Agency for International Development (USAID) provided significant support through technical assistance to help community organizations qualify and comply with new regulations for concessions. By 2005, ACOFOP had 22 member communities and organizations, representing 14,000 individuals in 30 communities. The members manage the largest expanse of forest under community concessions in the world, over 95 per cent of which has been certified by the Forest Stewardship Council (Nittler and Tschinkel, 2005).

The tenure rights granted to community organizations were broad and included extensive rights over the concessions. Nonetheless, their management rights were conditional on state approval. State agencies required local organizations to register, develop management plans, follow technical norms and certify their timber management operations. In addition, community concession organizations were required to control the holders of customary withdrawal rights for NTFPs. Figure 3.2 illustrates the strengthening of the collective entity to govern the common resource base and shows the significant expansion of management rights, though these were highly regulated. The government retained alienation rights and all rights were granted for the limited duration of the concession.

Although the concessions have functioned successfully, in some cases unresolved conflict exists over historical withdrawal rights to *chicle* trees

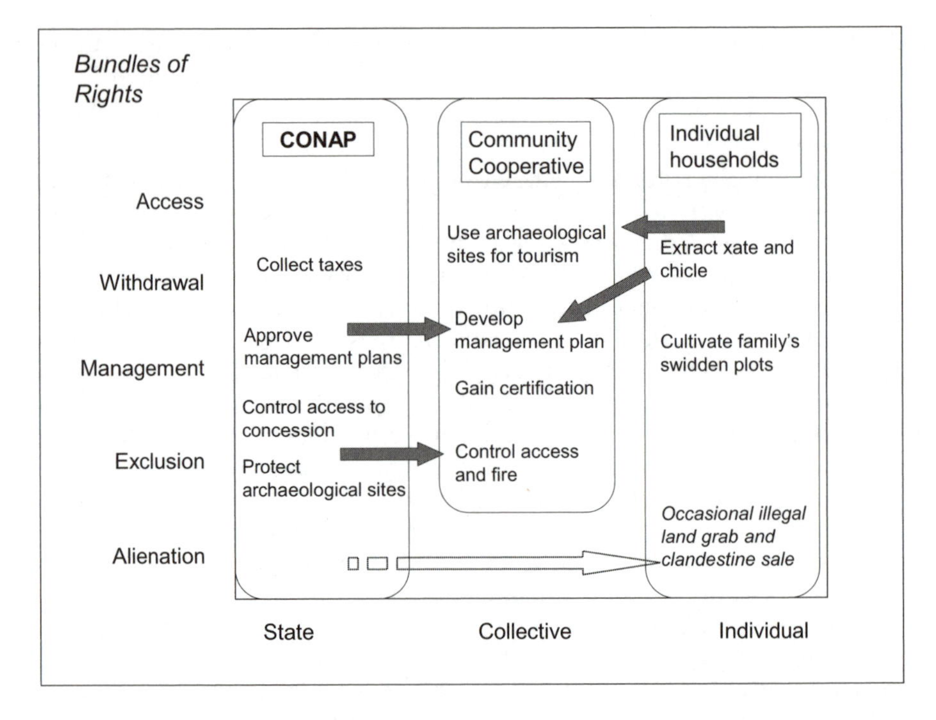

Source: prepared by author

Figure 3.2 *Community forest concessions in Petén*

and *xate* palm. The management rights granted to the community concession organizations have limited year-round NTFP harvests, an important stream of income for women and children. These collectors have no formal rights or representation in the concession organizations. Their discontent has been channelled to the leaders of the community cooperative, who are attempting to modify the regulations.

In general, the community concessions of the Petén do face significant obstacles, especially those operations with the weakest organizations and the least commercially valuable forests (Nittler and Tschinkel, 2005). In some cases, individuals have attempted to seize concession lands as private property. The organizations face scepticism and outright opposition from industry and some NGOs (Gómez and Méndez, 2005; Trópico Verde, 2005). Also, in a region with a growing population and significant landlessness, the concession organizations face a potential threat from groups questioning the allocation of such expansive areas to a small number of community organizations. CONAP is not the only government institution with influence in the Petén and the Maya Biosphere Reserve. Guatemala's tourism and cultural ministries, for example, have overlapping jurisdictions over resource management in the

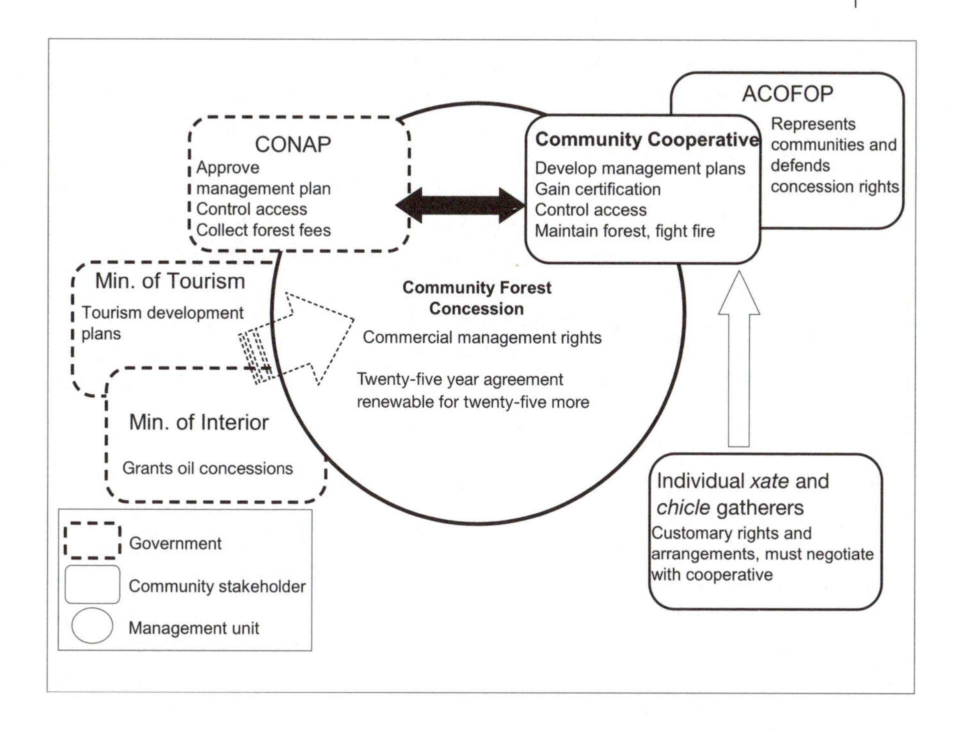

Source: prepared by author

Figure 3.3 *Co-management of Petén community concessions*

region. Suggested alternative policies have included efforts to create a new conservation area that would rescind forest concession rights. The government has also issued concessions for oil exploration that are superimposed on timber management areas.

Under the co-management system in the Petén (see Figure 3.3), transaction costs increased considerably not only in terms of the financial requirements to participate and gain approval, but also in the time it takes for communities to engage in these bureaucratic processes. For example, the requirements for management plans have to meet international standards for third-party certification. Compliance is complicated by the limited organizational and technical expertise of some community concession groups. Decision-making processes in communities can be slow and conflictive. The pace can make it difficult to react to change or respond to government requests.

Although management rights require compliance with exceedingly complex and cumbersome management rules, the sharing of rights and responsibilities with community concession groups has, in general, furthered forest conservation. In most community concessions, collective efforts were successful in diminishing illegal logging and archaeological looting. Community members have established their own local governance systems, based on an expanded set of rights of access, use and decision-making over their natural resources. This includes organizing patrols to protect concession boundaries and suppress fire.

Community organizations and ACOFOP have increased their capacity and strengthened their organizations. Government agencies, like CONAP, though weak and underfunded, provide help with legal procedures and some support in the field. However, significant investments were required from external funders to develop lumber mills and train communities in trade and certification standards – investments that allowed the community groups to convert new rights into livelihood and income improvements (Mollinedo et al, 2002).

India: Tree growers' cooperatives

The Tree Growers' Cooperative Society programme (TGCS) is a cooperative model created to establish and manage tree plantations on degraded village common lands. It is one of the clearest cases of tenure transfer for community-based forest management in India. It is also an interesting case because the management rights devolved to the cooperatives are relatively broad and a high share of the transaction costs are covered by the state. The cooperatives are provided with long-term leases to state-owned common lands (officially, 'revenue wasteland') for developing tree plantations and increasing fodder production. Decision-making power is devolved to local actors, but the leases are extended only to extremely small and degraded parcels.

The TGCS case study draws on village-level research conducted in the Ajmer district of Rajasthan state in northern India (Saigal et al, 2008). Rajasthan is the largest state in India, constituting 10 per cent of the country's area (GoI, 2008). Although 9 per cent of the state is classified as forestland, the actual forest cover is just 5 per cent (FSI, 2003). Most of the region (61 per cent) is either desert or semidesert (GoR, 2007) and as much as 30 per cent of the state is classified as wasteland (MoRD and NRSA, 2005). State lands classified as revenue wastelands are included in the wasteland total and are treated as common lands by villagers but are often *de facto* 'open access', causing further degradation.

The TGCS programme emerged in the 1980s from growing concern by Indian governmental agencies over fuel wood and fodder scarcity and increasing land degradation. The cooperative model for tree plantations was expected to be a promising institutional alternative to the existing social forestry programme and was launched with substantial funding from the Indian government as well as foreign donors (Saxena, 1996; NTGCF, 1997; Misra, 2002; IRMA, 2006). Programme activities were guided by an organization that became known as the National Tree Growers' Cooperative Federation Limited.

Under the TGCS model, villages were selected to organize and formally register cooperatives that could then lease government-owned wasteland. The leases could cover up to 40ha, were valid for 25 years and could be renewed

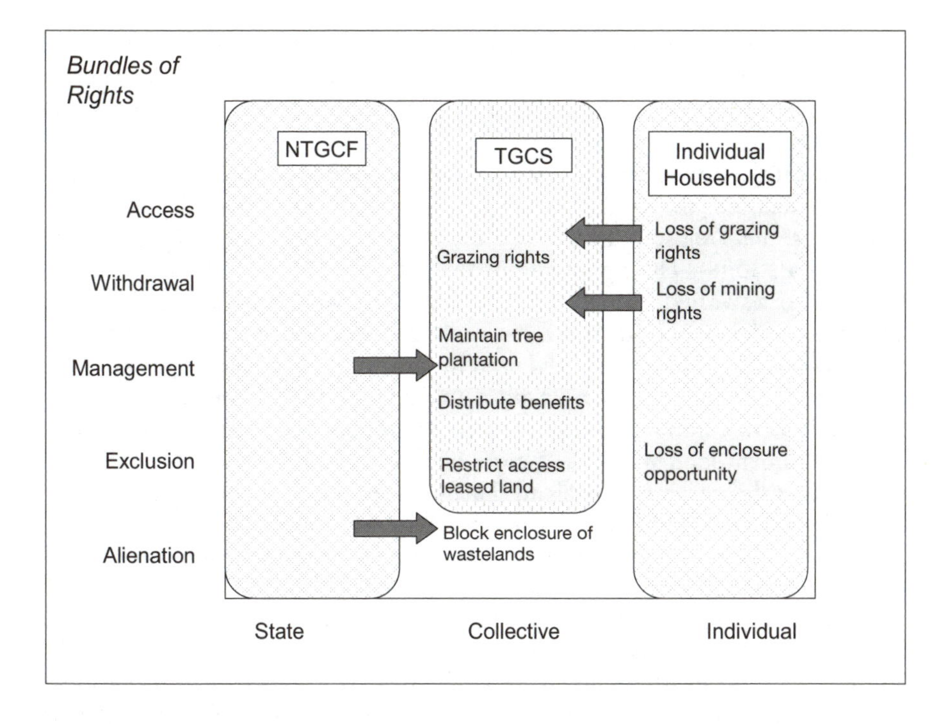

Source: prepared by author

Figure 3.4 *Devolution of wasteland rights to India's tree grower cooperatives*

for another 10, provided that there was no violation of the lease conditions. Membership in the cooperatives was drawn from the entire village with one member per household, each paying a nominal membership fee, although in practice other villagers might participate. The cooperatives established fuel wood and fodder plantations on the leased land and contracted guards to protect the plantations against illicit grazing, tree felling and collection of various forest products. The entire cost of the plantation was borne by the national federation, which for the first five years provided technical and programmatic support for field teams based in the region (NTGCF, 1997; IRMA, 2006; Singh, 2007). By 2007, there were 548 tree growers' cooperatives (FES, 2007). Financial support from Swedish and Canadian development agencies ended in 2001 and 2003 respectively (Singh, 2007) and at present there is no specific project supporting the cooperatives. Nonetheless, the leased concessions continue to operate.

The tenure reform represented by TGCS consisted of the temporary devolution of management and exclusion rights over state lands to cooperatives through leased concessions (see Figure 3.4). The cooperatives were granted management responsibility in return for investments, support from the National Tree Growers' Cooperative Federation Limited and the possibility of benefits.

Although expected to prevent encroachment and comply with rules related to cooperative administration, they were given considerable leeway in making management decisions, including how to harvest and distribute resources and how to organize management activities.

To illustrate how management decisions functioned in the cooperatives, it is helpful to examine the local context in greater detail. The three villages studied in this case had registered as TGCS organizations between 1991 and 1992 (Saigal et al, 2008). Their populations ranged from 82 to 220 households and their territories varied from 490 to 1716ha. There were broadly three types of land in the villages: private (mostly agriculture, both unirrigated and irrigated), village council land (common land used for grazing) and government land. The government revenue wasteland ranged from 31 to 262 ha and was generally held as *de facto* commons, used for grazing livestock and collecting fuel wood. The government land was in fact more like 'open access' areas, degraded and barren. It is this revenue wasteland that was leased to the cooperatives. Some of the leased land had been illegally privatized by other villagers and in all three cases TGCS removed illegal encroachments from the leased sites before starting the plantations.

In the co-management arrangement (see Figure 3.5), the leases temporarily provided the cooperatives with management and exclusion rights. The cooperatives received financial and technical support from the national federation and invested considerable effort and money to prepare the sites by carrying out soil and water conservation works, establishing fuel wood and fodder tree plantations, watering saplings and protecting the sites from illicit grazing and harvesting of tree products. Despite the investments, tree survival was low. The cooperatives controlled management decisions over various forest and tree products, including who, how and when to harvest. They closed the areas for several years to allow trees and grasses to grow and opened them for grazing only after trees were beyond browsing height. In effect, closing the areas restricted access rights to other community members who had claimed customary rights.

Although it has been more than ten years since external support ended, plantations in all three sites are still maintained and growing. The cooperatives are keen to renew their leases for the allotted land. The relatively secure tenure has encouraged members to invest in the land and protect the plantations, even after financial and technical support was withdrawn after five years. All three cooperatives kept their plantation guards after the project stopped paying them. Outside the TGCS concessions, the remaining village common lands are slowly being privatized.

The TGCS programme was intended to improve livelihood opportunities, increase fuel wood and fodder supplies and possibly generate cash income from the sale of tree products. Until recently, a major benefit was access to fodder in the post-monsoon period. A household had to pay the cooperative a fee to graze livestock, depending on the composition and size of its herd. The cooperatives used revenue from grazing fees to pay guards and the honorarium of the TGCS secretary. The main benefit at the moment is tree fodder (fresh

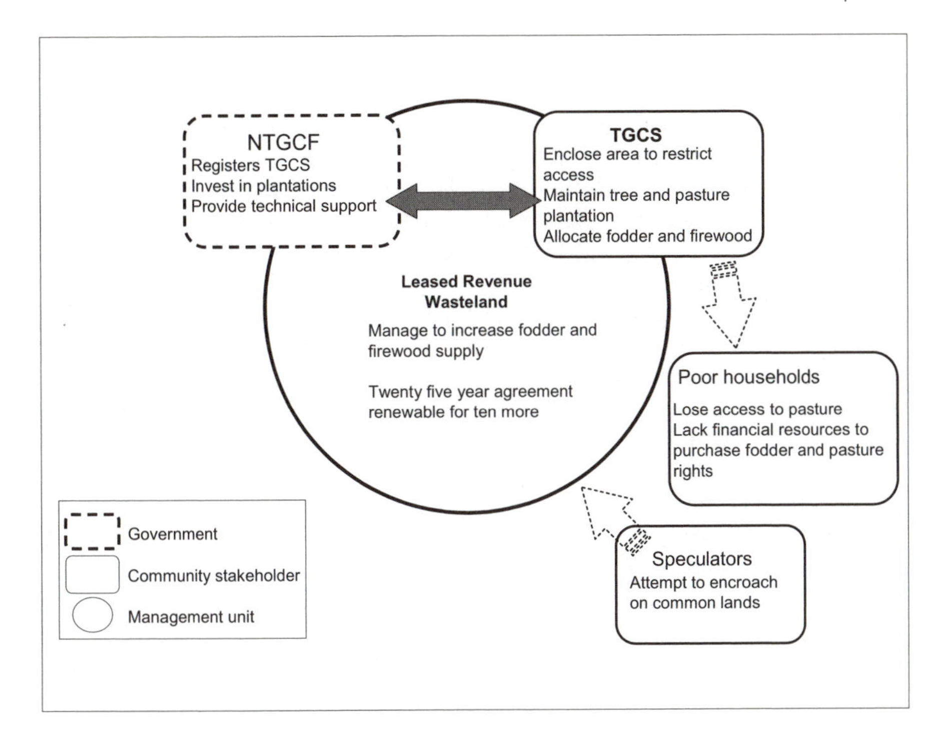

Source: prepared by author

Figure 3.5 *Co-management system in India*

leaves and pods) and some fuel wood. Supplies of fodder and fuel wood have increased only modestly. In the three study villages, a household gets on average only around 7 per cent of its fuel wood supply from the TGCS plantation; over half of the village's fuel wood comes from private lands. Considering that the benefits have been modest at best, it is not surprising that the majority of the respondents felt that the TGCS was at present 'unimportant' for their livelihood (Saigal et al, 2008).

One reason for the low impact is the 40ha limit on leases, regardless of population size. Such small parcels cannot generate substantial livelihood and income benefits for all village households. Furthermore, most leased lands were of poor quality and highly degraded when handed over to the cooperatives. The continuous drought over the past few years means there is hardly any fodder grass and the grazing fee system has been suspended. As a result, cash income has reduced sharply. The yields of fodder grasses have fallen in all three cooperatives, adversely affecting larger animals, such as cows and buffaloes. As a result, cooperatives have discontinued the practice of closing their plantation during the monsoon to allow grasses to regenerate.

The TGCS programme's impacts on village equity have been mixed. The differential impact on households is tied to the level of dependence on the plantation land. Households with more private land (particularly irrigated land) and thus a greater availability of fodder for livestock relied less on the commons. The poor have little private land and are most dependent on the commons. The concession programme inadvertently restricted access to the commons for the poor once the land was converted to a TGCS plantation. The closure of plantation areas for several years to allow trees to grow meant there were fewer alternative sources of pasture and fodder for poor families. The imposition of grazing fees seemed to be equitable, since the same rules applied to all; however, the poor had less ability to pay. Once the plantations had matured, the cooperatives decided to allocate fodder production through auctions, disadvantaging households of limited means. The stock of tree fodder and fuel wood from the entire plantation is auctioned to the highest bidder, who pays upfront and later resells the produce in smaller lots to others. This practice reduced the transaction costs for the cooperatives, but it clearly violated the principles of cooperation, since profits go to those who can pay while the poor are confronted by higher prices. The highest bidder need not even be a member of the cooperative.

The cooperatives were obligated to follow standard cooperative regulations, but because of weak oversight by the government the groups have gradually become less democratic. Elections for leaders are not held regularly and major decisions are made by a managing committee rather than the general assembly. More than just the state's failure to insist upon transparent and democratic procedures, the problem is also due to lack of awareness and apathy of members who remain passive. Residents mentioned that the cooperatives are dominated by particular caste groups (Saigal et al, 2008). The traditional marginalization of certain castes, the poor and women has also prevented broader participation in the organizations.

Despite these weaknesses, the TGCS programme seems to have resulted in better management of leased lands by giving local people a stake in managing resources. Outside the leased concessions there is widespread encroachment of common lands, partially motivated by past government policy that regularized encroached lands by titling them as private property. This is increasingly common where growing urbanization and industrialization are driving up land values. State policies encouraging investments in quarries to support economic development also put pressure on the village common lands. Yet there was no encroachment in TGCS plantations in two of the study villages and only minor encroachment in the third (less than one-third of a hectare).

Philippines: Forest resources development cooperative

The Ngan Panansalan Pagsabangan Forest Resources Development Cooperative is located in the Compostela Valley Province in the southern Philippines island of Mindanao. It represents a shift from a command-and-control system, in which the government granted forest resource rights to large-scale industry, to one where rights were transferred to community organizations to carry out forest management. The Ngan Panansalan Pagsabangan cooperative manages the second-largest community-based forest management project in the Philippines. It was the first community forestry operation in the Association of Southeast Asian Nations (ASEAN) region to earn SmartWood certification. Although the tenure rights offered innovative opportunities for community actors, the state has been slow to accommodate the new system and community groups like the cooperative face challenges because they lack the economic and political influence formerly held by timber industries in the region (Pulhin and Ramirez, 2008).

Forestland in the Philippines is owned by the state. The main agency responsible for its administration and management is the Department of Environment and Natural Resources (DENR). Until the mid-1990s the state used timber licensing agreements to award logging rights, usually favouring large timber companies. The companies were granted broad management rights over forests and claimed exclusion rights; the state retained alienation rights. There was frequent collusion between DENR and the companies and since DENR's monitoring was weak, the companies were often able to use their political and economic power to avoid supervision.

Prior to the 1990s forestry reform, the harvest of forest resources in Compostela Valley was controlled by the Valderrama Lumber Manufacturers Company Incorporated, which held the timber licence. The licence was for a 26,000 ha concession for 25 years, set to expire in 1994. Under the licensing agreement, the company was granted broad management decision powers but within an official regulatory framework. The government maintained control by monitoring forestry activities and requiring the development of annual operating plans defining the locations of cutting areas and the species and volumes to be cut. The government could suspend or terminate logging agreements if the company committed infractions. However, in practice, companies could avoid oversight by bribing DENR officials. Furthermore, to protect its operation, the company invested in forest law enforcement to exercise exclusion rights. Valderrama Lumber had its own security force to protect the forest; it filed criminal charges or evicted farmers who practiced swidden agriculture *(kaingin)* in the forest management area. During the peak of its operation, the company employed around 3,000 workers, mostly migrants. Mansaka-Mandaya indigenous people in the region were offered menial jobs hauling wood on a daily or contractual basis. When the licence expired, most of the migrants decided to remain.

While Valderrama Lumber held the licence, the Mansaka-Mandaya people maintained *de facto* access, withdrawal and exclusion rights based on customary claims but could not legally harvest or sell timber. They had their own institutions and customary rules governing forest use. Indigenous communities subdivided land for members and actively attempted to protect their forest from intrusion by other groups. Each family held small plots, usually about 10 ha, for swidden agriculture. These rights could be inherited by relatives and outsiders could gain access through intermarriage. Problems and conflicts were resolved by the council of elders. These informal institutions eventually clashed with formal rules used by the state to allocate forestlands. Some indigenous people practicing traditional farming systems began to be arrested by Valderrama Lumber and DENR. However, conflict diminished as the logging company offered jobs to indigenous people, who became more dependent on income from the company's timber extraction activities.

The denial of local peoples' legal rights to forest resources began to shift when, in 1989, DENR established the Community Forestry Program to provide upland farmers legal access to forests and financial benefits from these resources. Under the programme tenure was given to qualified community organizations through a community forest management agreement granted for a period of 25 years, renewable for another 25 years. In 1995, the Philippine government shifted the national strategy for sustainable forest management to the Community Based Forest Management Program. In this programme the government granted the right to occupy and manage certain forests and forestlands to community organizations by awarding forest management agreements. The big difference from the previous arrangement was that the government granted local organizations forest resource rights that previously had been held by industry. In other words, this entailed a shift from individual to collective rights. However, lacking the economic and political power of timber companies, community groups often struggled to enforce their rights. Also, because rights were limited to legally recognized local organizations, the reforms fell short of broad recognition of rights for individual local inhabitants. In fact the state's bureaucracy was unprepared for the paradigm shift and most of its old staff has had a hard time embracing its new function. Decision-making has remained centralized. Moreover, policies emanating from the national government tend to restrict rather than assist the cooperative.

Valderrama Lumber's timber licensing agreement expired at the end of 1994. With its departure, DENR became concerned that the forest would become an open access area without strong community organization to channel local use and benefits from forests. In response, DENR introduced the idea of forming a cooperative, with support from USAID. In 1995, a proposal was presented to local governments within the former concession area. Members of the Mansaka-Mandaya people initially did not agree with the idea, thinking that they would lose access to their landholdings. However, they eventually embraced it after DENR representatives carried out an educational campaign. The migrant workers who had stayed after the company pulled out were also potential cooperative members.

The Ngan Panansalan Pagsabangan Forest Resources Development Cooperative was formed and registered with the government in 1996. Later that same year, it was awarded a management agreement covering 14,800 ha of forestland outside the towns of Compostela and New Bataan. In 2004, this region had a total population of 8259 (approximately two persons per ha). The cooperative has 324 members, who together with their families represent a minority in the region (about 13 per cent of the total population). Former migrants make up 60 per cent of the membership and local Mansaka-Mandaya indigenous people the rest. Most indigenous people, therefore, continue to lack formal rights to forest resources.

Migrants formed the core group that took over forest management for the cooperative because of the skills they had developed while working for Valderrama Lumber. Day-to-day operations of the cooperative are primarily handled by experienced former employees. Representatives from the Mansaka-Mandaya people have participated in decision-making and occupy a majority of seats on the cooperative's board of directors. Some also work during the harvest as sawyers.

The cooperative's bundle of rights (see Figures 3.6 and 3.7) is basically the same as that held by the company. However, in practice, the situation is more contentious because of the government's oscillating policies regarding management rights and the cooperative's lack of political and economic power to exert influence on the government bureaucracy. Under community-based forest management, the cooperative is allowed to extract timber provided it develops a management plan (called a community resource management framework), prepares a medium-term plan projecting the timber volume to be harvested over five years and applies for an annual resource use permit.

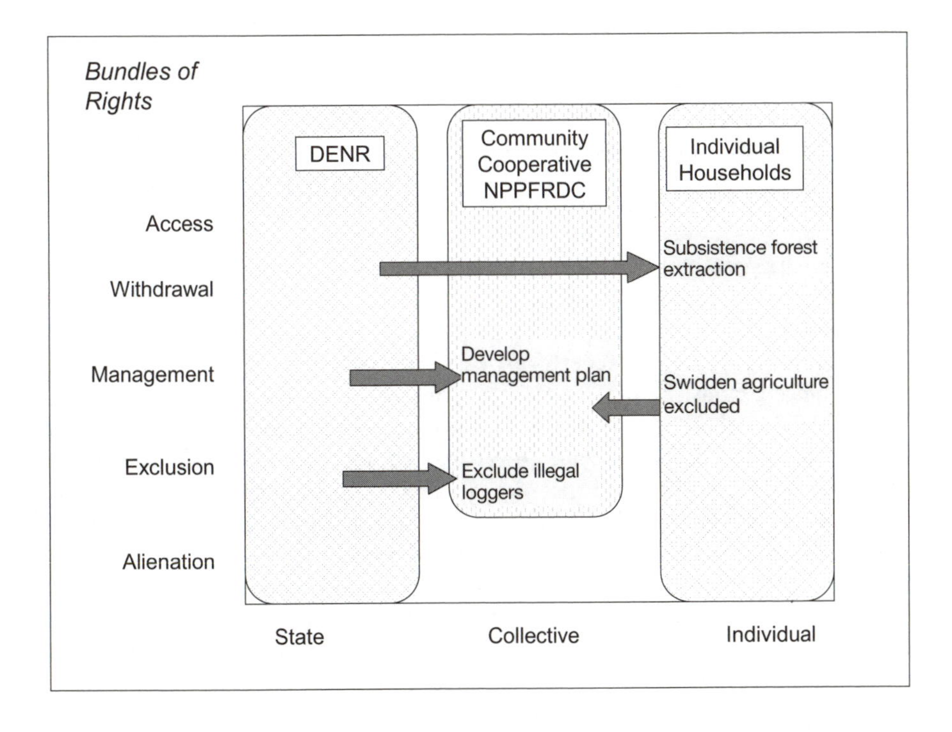

Source: prepared by author

Figure 3.6 *Forest property rights shift for communities in the Philippines*

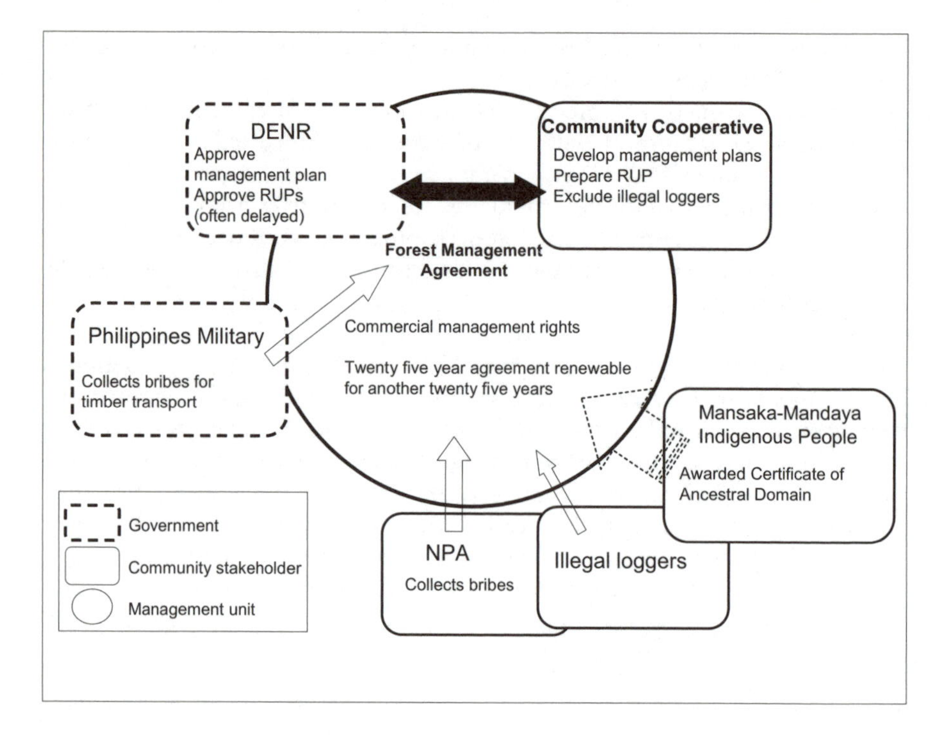

Source: prepared by author

Figure 3.7 *Co-management for communities in the Philippines*

However, the permit application process is tedious and entails high transaction costs: approval can easily take more than six months and costs almost US$5,000.

Despite the long and costly process, the permit is valid for only one year, counted from the end of the previous resource use permit operation. This has meant that, after delays, an approved permit is valid for less than six months. In addition, since 1998, DENR has issued three national suspensions of resource use permits. For instance, in 1998 DENR suspended all permits based on the allegations of abuse by a few community organizations and in 2006 DENR cancelled all existing community management agreements in eight regions, again because of charges of non-compliance or violations by community organizations (Pulhin, 2006). These suspensions applied even to well-run community organizations like the Ngan Panansalan Pagsabangan Forest Resources Development Cooperative, which was not given an exception despite its SmartWood certification.

The suspensions of resource use permits have eroded community members' motivation and commitment to protect and manage their forest (Guiang and Castillo, 2007). The uncertainty also takes a toll on the Ngan Panansalan

Pagsabangan cooperative's management and exclusion rights. Funds for forest development and protection depend on profits generated by the timber harvest and if delayed management activities suffer. Despite suspensions and delays in permit approval, DENR did not adjust the operational plans' development targets, placing greater pressure on the cooperative. In fact, the arrangement transfers most of the transaction costs and risks to the community cooperatives. Although the Ngan Panansalan Pagsabangan cooperative has suffered financially, it has sustained its activities without relying on external assistance. It has been able to innovate and make adjustments. On several occasions, it adopted a per board foot salary, paying workers based on the overall volume extracted per day multiplied by a given rate for each type of work. It has also implemented mass leave for all regular staff and workers whenever a resource use permit is suspended or awaiting approval.

Because the cooperative lacks the power and influence of the company, it also has to contend with other stakeholders who have moved into the power vacuum. Several groups now claim rights to resource use in the area, including the military and the New People's Army, a rebel group that considers the forest management area its base. From time to time, the cooperative must pay bribes to the military, the rebel group and DENR, just to secure the safe passage of their timber products and avoid delays in the transportation of logs. Because the cooperative does not have the money to hire security guards, illegal loggers have been drawn to the site. These illegal loggers harass the cooperative's staff and in some instances have used threats of violence to intimidate the cooperative and assert their *de facto* withdrawal rights.

Many residents see little difference between the cooperative and Valderrama Lumber. This may be true at the organizational level, since many former employees are members. However, the cooperative lacks the company's financial resources. In addition to the problems posed by policies of the national government, it experiences difficulties in complying with its obligations for forest development, rehabilitation, timber extraction and forest protection. The tenure rights of Mansaka-Mandaya indigenous people improved as their organization, the Kaimunan ng Lumad Compostela, was awarded a certificate of ancestral domain title over an extensive area that includes the concession controlled by the cooperative. Once the cooperative's 25-year concession expires, it will need to negotiate forest management rights with the indigenous organization that holds title, and it is not clear how this will work.

Bolivia: Indigenous rights in lowland communities

The Bolivia case study provides an example of forest tenure reform that devolved management rights to indigenous communities, opening opportunity for commercial management of forest resources. For some, the changes have generated substantial benefits; for others, the regulatory framework and difficulties in complying with conditions set by the state agencies have limited their chances to participate.

A series of reforms in the 1990s changed the bundle of tenure rights available to Bolivia's indigenous people. The tenure reform law (known as the 'INRA law' for the institution it created, *Instituto Nacional de Reforma Agraria,* National Institute of Agrarian Reform) recognized a type of communal property called a TCO (*tierra comunitaria de origen,* original community land) that offered rights and decision-making powers to indigenous populations. At the same time, a new forestry law transferred management rights to indigenous people within TCOs. These reforms opened up forest management opportunities that could provide indigenous people with new sources of income. But as our case study illustrates, onerous regulatory frameworks and inefficiencies created bottlenecks in the co-management arrangements with governmental agencies. As a result, benefits have not been as extensive as anticipated and some indigenous people interested in commercial forest management have been unable to participate.

To examine how the devolution of forest management rights affected indigenous people, the case study considers Guarayos communities in Bolivia's Santa Cruz department (Cronkleton et al, 2009). The Guarayos TCO covers much of the Guarayos province's 29,433km^2. At the time of the TCO's creation, the region had only 31,577 inhabitants (INE, 2001). However, the construction of an interdepartmental highway has opened the region to outsiders, leading to a population increase. The influx has produced an ethnically mixed population and the indigenous population struggles to maintain the security of its property.

Prior to the reforms, most indigenous people in the region lacked formally recognized property rights. Lands and forests around villages were controlled as *de facto* communal lands, with households granted individual rights to cultivate small plots. The state owned forest resources and, technically, much of the land in the region. Some rights were allocated to private property owners or industries but generally not to indigenous people. Indigenous people had *de facto* control over their agricultural plots and struggled to defend communal lands around villages, where they informally claimed access and withdrawal rights for subsistence purposes. Indigenous people were denied formal rights over forests; the state instead granted management rights to a select group of timber companies through long-term contracts giving them exclusive rights to forests resources. These contracts superseded competing property claims of communities or private property owners and indigenous farmers were denied the ability to exclude timber companies from their lands.

The 1996 INRA and forestry laws transferred rights to a broader range of stakeholders and shifted the bundle of rights available to indigenous people (see Figure 3.8). The new rights were collective within TCOs. As defined by the INRA law, a TCO is a communal property that covers lands traditionally occupied and used by the indigenous populations. TCOs are inalienable, indivisible, non-reversible, collective and non-mortgageable as well as tax exempt. Within these communal properties, the internal distribution and use of resources, access and withdrawal rights are determined by residents' *usos y costumbres* (uses and customs), although they are still required to follow agrarian and forestry regulations. Forest management rights for subsistence resource use were devolved almost completely, again defined by customary

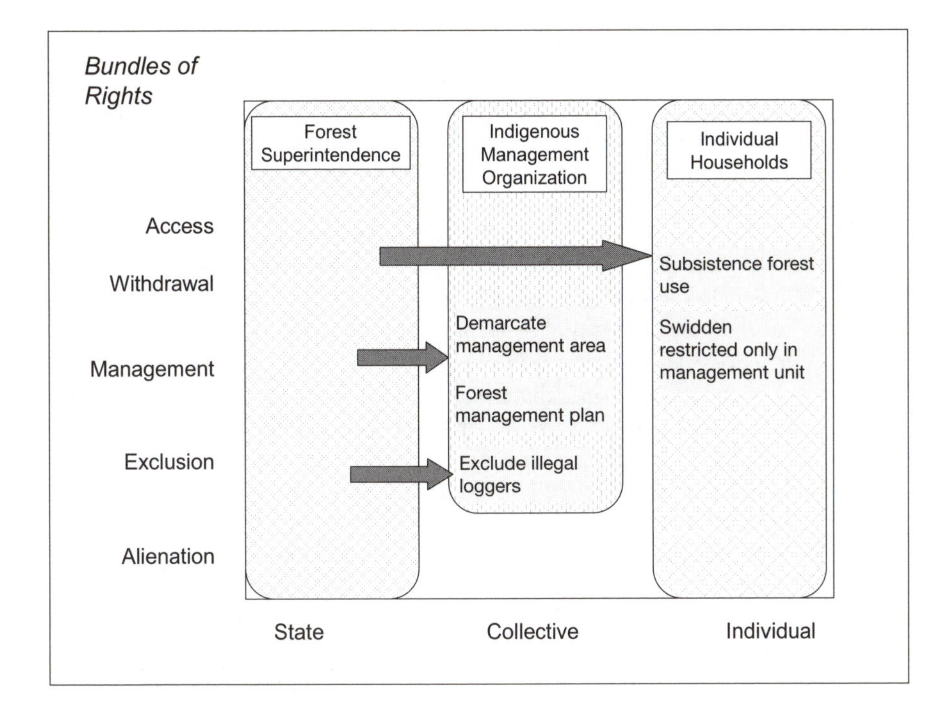

Source: prepared by author

Figure 3.8 *Forest tenure rights shift in Bolivia*

practice and with little involvement by state agencies. However, for the commercial use of forest resources, management rights were allocated only with approval from the state's Forest Superintendence.

Gaining approval for management rights from the state was a time-consuming and costly process. For approval of management rights, indigenous people needed to form management organizations and document their uncontested control over the designated forest management unit with approval from the Guarayos indigenous organization (COPNAG). They also needed to carry out an inventory and develop a management plan that conformed to the government's technical norms. Reaching the approval stage could take months or years and cost thousands of dollars. After approval, the indigenous organization had to submit an annual operating plan to be permitted to harvest and then submit an annual harvest report.

Although the forestry law was intended to promote sustainable forest management, the legislation also recognized that rights holders would harvest timber as part of forest conversion for agricultural purposes. Therefore, exceptions were included through mechanisms that provided one-time authorizations for the sale of timber from agricultural areas, amounting to commercial withdrawal rights. Technical norms were created for logging permits during agricultural conversion and other small-scale operations that were administratively simpler and not subject to the same rigorous enforcement as general forest management plans. These alternatives were not supposed to play a major role in the sector, but as indigenous people encountered bottlenecks to forest management, they turned to less sustainable alternatives like forest conversion.

When the Guarayos TCO demand was presented in 1996, the government determined that the property should cover 1.3 million ha (VAIPO, 1999). Titling has been slow and after a decade it is still not complete, particularly in areas near the highway with higher population densities and more frequently contested property claims. Although progress in remote areas has been rapid, this was little consolation for indigenous people who do not have their lands defined and are under pressure from other stakeholders. In response to the slow process, Guarayos groups embraced community forest management as an alternative that would help them control communal forests and provide another opportunity to generate income.

To gain management rights, indigenous communities in the Guarayos TCO had to seek assistance from NGOs that could provide technical support and subsidize the costs of preparing management plans. Although the TCO is supposedly a huge, indivisible communal property, management units were developed at the village scale. To date, seven communities have gained approval for general forest management plans in Guarayos covering some 150,000ha of forest. The management plans benefit approximately 250 indigenous households directly with wage labour and profits from timber sales. Even though timber companies had already removed the high-grade timber, significant volumes of alternative species with commercial value remain. When these communities harvest and sell timber, they can potentially generate tens of thousands of dollars in gross income.

The co-management arrangements (see Figure 3.9) with the government have created disincentives for the community forest management groups. The communities bear a high proportion of the transaction costs and risks of the system, and the government agencies have not fully met their responsibilities for supporting community rights and restricting illegal and unsustainable timber management. For example, of the seven community forest management projects, three have struggled with encroachment by outsiders. Requests for assistance from INRA and the Forest Superintendence to defend their management units and reaffirm their exclusion rights have gone unanswered. Because of the slow titling process, it has been difficult for some communities to control uncontested forest areas near their settlements. One of the original seven community groups found that a large portion of their forest management unit had been claimed by a private landowner. The community fought to keep their plan but eventually abandoned their effort. Two other CFM projects have had to continuously battle encroachment on their forest management unit by colonists and ranchers.

Indigenous people outside these groups or in other communities have found it difficult to take advantage of the management rights. They needed

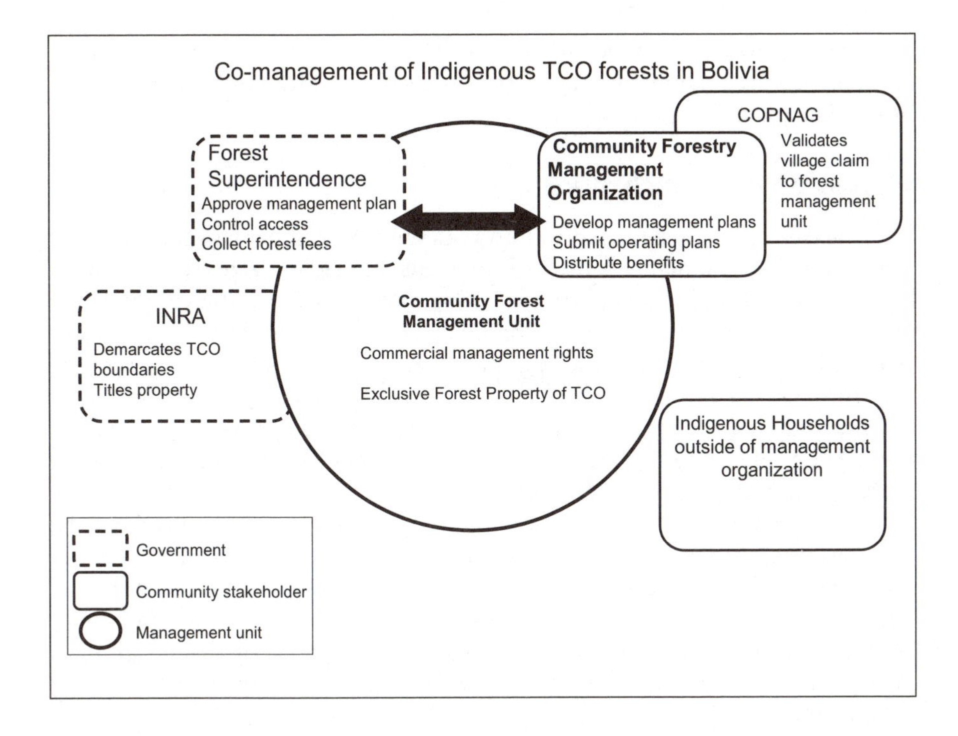

Source: prepared by author

Figure 3.9 *Co-management in Bolivia*

uncontested forests, which were not always available. Furthermore, NGOs promoted forest management in large areas of high-quality forests, yet many indigenous families live in areas with fragmented and degraded forests and could not benefit from the NGO subsidies and assistance. As a result, increasing numbers of indigenous households have relied on alternative mechanisms – land clearing and logging permits – that provide one-time withdrawal rights but are less sustainable.

Exercising all of their rights has not proved as easy as expected for Guarayos families. The right of exclusion required that the government define the area over which the right applied. The titling process has been long and drawn out and the area claimed by the Guarayos people was ethnically mixed, meaning that outsiders and other competing stakeholders were already within the claimed area. The state had to determine which competing claims would be recognized and only then, when contested claims over areas were resolved, would the area be turned over to the TCO residents. The size of the territory limited the ability of indigenous people to enforce their right of exclusion and their claims were not backed up by state agencies or courts. As the state advances with the titling, it has given preference to private land claims and even given priority to timber industries. The government's response to land invasion has been ineffective and illegal land sales and fraudulent transactions have taken place.

Discussion and conclusions

The four cases presented here show how community forest management projects were shaped by the manner in which management rights were transferred. In these cases the transfer was partial, creating management arrangements where responsibilities and decisions were shared by government and local stakeholder groups. Even though tenure rights granted to communities allowed customary decision-making mechanisms to guide subsistence management activities, commercial management was different, and state agencies retained major roles. The tendency to give rights and then take them back through regulations illustrates not only a tension between conservation and development agendas, but also a lack of confidence in community stakeholders. It also reflects the fact that the poor could be vulnerable to manipulation by outsiders without government oversight. The controls are intended to stop illegal resource use, but the cost of these measures often falls disproportionately on the community groups attempting to comply with the law. Illegal loggers, by definition, have incentives to avoid governmental controls, while community groups must demonstrate compliance.

What management rights were transferred and which were retained by the state? In all four cases, the state retained ownership of the forests (i.e. alienation rights) and offered some management rights and responsibilities but always retained a role in the process. In each case, the ability to exercise management rights required that community-level stakeholders carry out specific tasks to gain authorization. The state maintained control of management by requiring compliance with planning formats, technical standards and organizational models and administrative procedures. The granting of management rights was conditional and, in three cases, temporary. Non-compliance with the system could mean that local user groups would lose their rights. Nonetheless, within these co-management systems, significant power and responsibility were transferred to local stakeholders. All the groups could and did make operational decisions about their forest use; however, in all cases they were required to stay within established parameters.

In all four cases the regulatory framework guiding the community forest management models created barriers to participation, such as high technical standards, costly and time-consuming procedures and other requirements. Although the state offered co-management opportunities, these were beyond the reach of many households that lacked capacity and access to subsidized support. In only one case (India) did the state take most responsibility for the transaction costs, but this was for small, degraded areas. In all cases, the community groups required external assistance to take advantage of new rights to forest resources. Usually, assistance allowed them to comply with technical standards related to sustainability. After passing the hurdle of high startup costs, the community groups could usually maintain their role within the co-management arrangements, although they frequently confronted obstacles from regulators.

Who was ultimately granted the management right? In these four cases, governmental agencies required local people to form specific types of organizations to take responsibility for collective rights. Community stakeholders could make decisions about who participated and how. However, they frequently adopted standardized models introduced by the NGOs subsidizing the startup costs. By insisting on the formation of local organizations, the government could be assured that a specific entity (and individuals) would be responsible for management actions. It was also assured that the necessary institutional arrangements were present before it devolved the rights. One unintended consequence of the formation of management organizations is that these new organizations could overlay previous existing organizations (see Chapter 4), displacing or excluding other stakeholders. In India, those who were excluded were often the poor and most dependent on the resource.

In general, the communities received forests that had already been logged or were otherwise degraded. However, the forests still had commercial value that made the investment in management practices worthwhile. The degradation was an incentive for the government to extend the possibility of management rights and introduce co-management schemes. In all four cases, the community forest management models were successful in improving forest governance and decreasing destructive practices.

The partial devolution of management rights through co-management systems had several trade-offs:

- Giving up some legal control over resource use decisions allowed the government to gain greater cooperation from local stakeholders. Because community-level stakeholders gain benefits by complying with official rules and guidelines, they are more likely to have vested interests in maintaining forest resources.
- The increase in benefits for community groups comes with increased responsibility and obligation. In some cases, the burden of new regulations and intrusion by government outweighs the benefits of participation for some local people.
- Devolving rights to some groups means that other groups and individuals are excluded. In fact, if there is no preexisting competition over the resource, granting exclusive rights is rather meaningless.
- Co-management is a process, and monitoring and reflection are required to continually update and improve the system. State forest bureaucracies are often rigid or unwilling to evaluate or adjust regulatory frameworks, but flexibility is necessary. Balancing the need to address local contexts without dissipating policy frameworks into myriad locally specific rules is a challenge.
- The temporal nature of some rights transfers allows the state to maintain control but creates tenure insecurity. Because the state can rescind the

management rights, long-term investments based on the assumption of continued benefit streams can be risky.

A central finding is that the partial retention of management rights by the state creates persistent, significant barriers to the adoption of community forestry and in some cases it has limited the benefits to local participants. In the worst cases, state regulations introduced to guide decisions become rigid frameworks that limit the kinds of choices that can be made, which in turn inhibits adaptation. Mechanisms are needed to facilitate dialogue between state agencies and communities to consider more local input into the design and revision of regulations.

4

From Discourse to Policy: The Practical Interface of Statutory and Customary Land and Forest Rights

Emmanuel Marfo, Carol J. Pierce Colfer,
Bocar Kante and Silvel Elías

We drove along the winding road, surrounded by grassy hills in the foreground, forests in the distance, struck by the poignancy of the situation. Many in the Lao-PDR government were trying to manage the area in such a way as to improve people's livelihoods. But formal government policy was tied to an 18th century political philosophy of human social evolution that hid from them the complex agroforestry and governance systems which local peoples had evolved to organize and sustain themselves over the millennia. The government's forest reform involved managing the forests and communities across the land in a uniform way, designed from an office in the capitol – an approach deemed easiest to implement. The policies involved uprooting peoples from remote areas where their forest gardens flourish with mature trees from which they harvest fruits, bark and wood, and around which carefully tended swiddens and gardens provide food, medicines, spices and herbs for their families and their community. Resettlement would establish family clusters along the road, in small areas of land according to pre-set formulae. Age old forms of customary practice, leadership and laws were cast aside, as the government ignored the wealth of cultural diversity and appropriate local governance, resulting in an uprooted people, a damaged landscape, sad disempowerment.
(Colfer's personal observations on the Land and Forest Allocation Program, March 2008)

Land and forest policies in many countries, including Lao-PDR, are formalizing recognition of local peoples' rights, especially to provide them with opportunities that improve their livelihoods. However, as observed by Cousins (2007b, p291), 'these policies must take cognisance of the complexities and realities of current regimes of claims, rights and their governance, i.e. how "actually-existing" tenure systems operate in practice'. Customary tenure systems, by definition,[1] have evolved over long periods of time in response to location-specific conditions (World Bank, 2003). In the process of recognition, such customary systems have been ignored, subordinated or, at times, effectively accommodated (Tahamana, 2007; see also Elbow et al, 1998, cited in Diaw, 2005).

The scholarly debate on whether to accept one legal system as superior to others, or what their respective weights should be, continues. There is a call for a paradigm shift from legal pluralism, which recognizes parallel systems, to legal integration, which would mesh them. Integration would require understanding the major constitutive elements of each system (Diaw, personal communication), in each national context, as well as the ways in which they have been accommodated or subverted, in order to design an effective negotiation that aims at combining their strengths. It would require maintenance of the widely acknowledged flexibility and responsiveness characteristic of customary systems. In the context of this book, the goal would be to better secure forest tenure rights, especially for marginalized peoples.

Tenure norms and rights are embedded in the institutions and laws that govern the entire sphere of human activity. Their integration, therefore, is far more complex than can be addressed by the consideration of tenure rights alone. Although an in-depth proposal for legal integration is beyond the scope of this chapter, it is important to understand how statutory and customary laws within specific socio-political settings interplay and have coexisted. Based on studies in Ghana, Burkina Faso, Indonesia and the highlands in Guatemala, this chapter examines the relationship between statutory and customary land and forest tenure, the models by which customary laws have been recognized and the extent to which statutory law has accommodated or subverted customary systems.

The chapter proceeds with a brief elaboration of the conceptual issues related to the interface between customary and formal tenure regimes. Descriptions of the four cases follow. Finally, the cases are briefly synthesized to draw out the cross-cutting lessons and highlight some challenges in the move from legal pluralism to legal integration.

Tenure rights as a legally pluralistic phenomenon

'In today's world, constructions of rights are conspicuously rooted in normative schemes generated by the institutions of the state and refined and elaborated by the doctrines of legal and political science' (von Benda-Beckman, 1997, p1). These state-generated normative schemes have come to be conventionally referred to as law. The literature on the law–state nexus is rich, with law deriving from notions of sovereignty and the state's monopoly of legitimate use of power and violence (see Faulks, 1999; Weber, 1968). The state's ultimate legal authority is based on 'the normative notions of internal and external sovereignty which comprise the state's authority to exercise exclusive control over the population that inhabits a territory and the wealth and resources that exist within the territory' (von Benda-Beckman, 1997, p4).

Nevertheless, in contemporary human societies, one often finds other substate political organizations with their own normative constructions, which may contradict those of the state.[2] Such normative orders may be based on so-called folk, customary or religious systems (von Benda-Beckman, 1997; von Benda-Beckman and von Benda-Beckman, 2002). Von Benda Beckman (1997) observed that 'where state law officially makes matters of social and ethnic origin irrelevant and allocates economic and political rights and duties on the basis of abstract equality, village law may do just the contrary'. He emphasized that whether one is seen as having the right of participation in decision-making or access to natural resources, as a citizen or a stranger, depends on the specific normative construction chosen.

Diaw (2005) has made a start at defining important characteristics of customary systems of land and forest tenure in Africa, features that are also important in many Asian and Latin American sites. Recognizing site variability, he identifies the following four features of land tenure among forest-dwelling groups:

1 'collective property', or the territories held or claimed by groups of people (e.g. communities, tribes);
2 'open access' property, where anyone passing through can harvest plants or animals at will;
3 'common property', in which products or areas are open to group members but denied to outsiders without special permission; and
4 'private holdings', portions of the 'common property' that are managed and in some sense 'owned' by individuals or families.

These are hierarchical, in that there are typically both open access and common property areas within a group's territory, and private holdings tend to be drawn from common property. Another crucial element of such customary systems is their basis in kinship, which often defines group membership (Agbosu, 2000; Diaw, 2005).

So long as these other non-state normative orders are used as legitimate resources in social conflicts and claims, as observed by von Benda-Beckman (1997, p6), 'some construction of the interrelationships between these systems becomes necessary, in which the respective scope of validity of the systems and their position in the political organisation is circumscribed'. The definition of this relationship creates a hierarchy of laws, usually acknowledging one's normative order as 'dominant' and interpreting the others as subject to it. The analytical notion that has been used to explore the social practice of law is legal pluralism, defined as the coexistence and interaction of multiple legal orders

(Meinzen-Dick and Pradhan, 2001) or different legal mechanisms applicable to the same situation (Vanderlinden, 1989).

These situations create conflicts and multiple sources of legitimizing claims over tenure rights, leading to 'forum shopping and shopping forums' (von Benda-Beckmann, 1981), where actors seek and use the legal system that best supports their interests. In natural resource management specifically, it leads to what Onibon et al (1999) call a 'sterile dualism', whereby the state imposes laws and regulations that are simply impractical and incompatible with local practices – hence the rules are simply ignored, while local people's behaviour is criminalized (Benjamin, 2008).

Alden Wily (2008, p46) argues that 'pursuance of statutory or customary legal regimes is not an either/or' proposition. Rather, there is a dynamic interplay among 'state authority, local power relations and inter-group resource competition' (Fitzpatrick, 2005, p454). Alden Wily (2008, p46) further observes:

> *...the customary rights of the majority, including common property rights, depend profoundly upon the support of statutes – i.e. national or state laws deriving from acts of elected parliaments. Assurance that customary regimes may operate in designated spheres and that the rights they deliver will be upheld as private property rights needs constitutional or at the very least modern land law support.*

Hence the call for integration.[3]

In the four cases discussed below, customary law is given some form of recognition by the state. Ghana's constitution recognizes customary law as one of the laws of the land. In Burkina, though the Land Act does not give explicit recognition, the constitution acknowledges customary rights. In Guatemala, the state pledged to recognize customary rights in the Peace Accords in 1996. Indonesia is the least clear on this issue, particularly with regard to forestland – though there *is* recognition of the existence and importance of *adat* (custom). Legal pluralism is a social reality in these contexts, but its practice has been characterized by a variety of problems and conflicts.

Before moving towards integration, it is important to understand the social practice of land and forest tenure laws and their interaction. This chapter uses two types of analysis. First, Tahamana (2007), cited in Benjamin (2008), has identified three main ways in which states typically approach competing normative orders, such as customary institutions: ambivalence, absorption and suppression.[4] The first refers to taking a neutral position or no formal repressive action; the second refers to recognition and support, and the third refers to aggressive attempts to repress or replace customary systems. Though all the cases studied here involve recognition of customary systems to some degree, the overall intention or goals of the state may not be to accept, support or integrate customary systems.

Second, Fitzpatrick (2005) presents four models by which customary lands have been recognized, each implying increasing degrees of state intervention. He argues that the central factor in defining which model to use should be the source of tenure insecurity affecting the community. He also cautions that these models assume a relatively benign state (an assumption that is not always justified in forested contexts). Each is discussed in turn. (It has not been possible to maintain the distinction between land and forest tenure, as described in Chapter 2, because customary systems often refer to both simultaneously.)

The *minimalist method* is the situation where the state merely recognizes customary lands and local norms, establishes a land registry and does not intervene in the internal matters of the community. This would involve, for example, drawing lines on a map and, at most, protecting the borders from outside intervention. According to Fitzpatrick, this model would be most appropriate when threats to security arise from outside the group. This model has few disadvantages, though the failure to promote any kind of integration with the rest of society or formal legal systems is likely to have some drawbacks except in the increasingly rare cases of communities that have had little or no contact with outside societies.

The second is the *agency method,* where state intervention is limited to identifying an agent who then represents the customary group. This was a common colonial solution to dealing with local groups, used in British colonial Africa for example. This approach simplifies matters for the state: it deals with the representative only and plays no role in the internal affairs of the group. This model, however, has the serious disadvantage of empowering the agents, who then may not act in the interests of the group or be accountable to the group for their decisions (see Ribot, 1999; Marfo, 2004; Oyono et al, 2008; Ribot et al, 2008). Today, no sub-Saharan African countries continue to use pure agency models and most are moving towards the land board model, described below (Fitzpatrick, 2005).

The third method is *group incorporation,* whereby a customary group incorporates into a recognized entity with legal standing, such as a cooperative, enterprise or other corporate structure. This involves the writing of by-laws and internal rules of procedure that help guarantee clarity regarding decision-making, recourse for inappropriate behaviour and conflict resolution. Fitzpatrick argues that this model is particularly useful if a community is to enter into agreements (such as logging contracts) with outsiders because it guarantees the legality of such agreements before the state. It also provides a mechanism for limiting the power of the group's leaders and ensuring more equitable decision-making within the group. This model raises other issues, however. In the cases studied in this research, for example, the models available to communities sometimes differed from the ways in which they were used to organize or make decisions (see Larson and Mendoza-Lewis, 2009; Larson et al, 2008). In the francophone areas of Cameroon and in Indonesian villages, such corporate entities may totally bypass customary institutions (Diaw, personal communication, 2009; see also Oyono et al, 2008).

A fourth approach to recognizing and managing customary tenure, adopted by some countries in Africa, is to establish a decentralized system of *land boards*. Here, the authority of traditional leaders is transferred by law to boards, which typically include both elected and appointed members, sometimes with traditional leaders as *ex officio* members. The boards are charged with holding the land in trust for the benefit of local communities or tribespeople, as well as for outsiders. This model represents the greatest degree of state intervention and includes characteristics that deal more effectively with outsiders' interests in customary lands than the previous types. The risks, however, include failure of the boards to allocate rights fairly and be accountable and the possible breakdown of customary systems (Fitzpatrick, 2005). The land board model also includes village councils, which are similar to boards except that members are elected locally; in Burkina Faso, such boards remain under the authority of local government, which has the decision-making power.

Each country discussed below was colonized by a different colonial power: Ghana by England, Burkina Faso by France, Indonesia by The Netherlands, and Guatemala by Spain. Although this implies different patterns, the similarities in forested contexts are perhaps more striking. In all four cases, there is a history of colonial usurpation of customary forests, beginning with their classification as empty or vacant and thus available for state or private colonial uses. Unsophisticated forest peoples were often tricked by the state and private companies (see for example Agbosu, 2000 on Ghana), a process that continues to this day in Indonesia (Colchester et al, 2006). None of the four colonial powers recognized the importance of swidden fallows or the existence of complex customary systems. And all gave primacy to private lands, ideally proven by title. The remains of this colonial pattern remain visible, to a varying degree, in the policies of independent nations.

Here, for Ghana, Burkina Faso, Indonesia and the Guatemalan Highlands, we discuss the definition of tenure rights by statutory and customary laws and the ways in which statutory systems have accommodated or subverted customary systems.

Ghana

In Ghana, the highest statutory law, the constitution, recognizes customary law as a legitimate legal order. Article 11(3) states that '"customary law" means the rules of law which by custom are applicable to particular communities in Ghana'. The recognition of customary law gives Ghana dual legal political entities, where issues of rights can be contested on the basis of both statutory and customary law. This by no means suggests that Ghana has a well-codified system of customary laws, however. In fact, the content and meaning of customary laws have often been disputed, and in these cases, 'the courts have relied on witnesses acquainted with native customs until particular customs, by frequent proof in the courts, become so notorious that the courts take judicial notice of them' (Woodman, 1996, p40). The courts have also looked to previous trends as authoritative precedents. Thus, J. A. Sowah, quoted by

Woodman (1996, p43), held that 'Whatever be the content of a custom, if it becomes an issue in litigation and the courts are invited to pronounce thereon, any declaration made by the courts supersedes the custom however ancient and becomes law obligatory upon those who come within its confines.'

Therefore, despite the formal recognition of customary laws, the content of custom, when in doubt, is taken as a question of law and not of fact. Boni (2005, p9), writing about Ghana, observed that 'while legal studies examine land tenure as "traditional" and therefore largely static or not subject to legislative innovations, land rights practices have in fact been subject to profound alterations and have undergone a continuous process of redefinition'.

Suffice it to say that both the courts and the customary system in Ghana have elegantly evolved to deal with land-related issues using interpretations of the same customary law.

Statutory recognition of customary land tenure rights

Community lands in Ghana are held by various stools,[5] or families or clans. The highest title in land recognized by law is known as the allodial title and in many traditional areas is acknowledged as being vested in their stools only. Hence the occupants of these stools are usually referred to as landowners. The only way one can acquire the allodial title is by discovery, that is, as the first hunter who identified the land and by subsequent settlement and use. The state, however, can acquire lands (which become public lands) from traditional allodial holders in two main ways. First, it can acquire land through compulsory acquisition in the public interest (Act 125), in which case all previous interests are extinguished; both the legal and beneficial titles are vested in the president, and lump-sum compensation is paid to the victims of expropriation (Kasanga and Kotey, 2001). Second, it can acquire land that has been vested in the president ('vested lands'), in trust for a landholding community (Act 123). In this case the legal title is transferred to the state but beneficial interests still rest with the community; here the government does not pay any compensation.

The customary right of ownership has been observed by the state since colonial days, when permanent forest reserves were created (see Agbosu, 2000). Land and forests have continued to be the property of the community even while the government manages them for the collective good. All benefit-sharing arrangements for revenue accruing from forest exploitation, defined in the 1992 Constitution, are based on this principle: lands belong to communities, and even if they are vested lands, communities are still entitled to benefits.

The state also recognizes the customary freehold. As a right and by virtue of membership in a community with allodial title, individuals hold a customary freehold to a portion of the land that they cultivate first or that is allotted to them by the community. The holder has the right of occupation, which may be passed down to successors (Da Rocha and Lodoh, 1999). Many native peoples in forest-fringe communities have such a customary freehold interest in their farmlands that has been passed on from their ancestors. This right is recognized and statutory laws on timber rights allocation require that farmers and landowners be consulted and their consent obtained before any timber operation can take place on their land (Legal Instrument [LI] 1949). Even during the creation of forest reserves, portions that were under settlement or cultivation were demarcated as 'admitted farms', and these lands have been cultivated to this day.

All formal titling must follow statutory procedures, which require evidence of possession of the specific right. This evidence is usually in the form of an allocation letter provided by the original rights holder, usually by the community or stool land chief.

State recognition

The situation in Ghana fits squarely into Fitzpatrick's agency model of state recognition, since statutory law recognizes chiefs as traditional authorities who represent community interests exercising their customary rights in land. This representation is complex: within the community traditional system are multiple layers of authority that claim control and hold some rights or interests over specific community lands (see Chapter 5). For example, even though the allodial title over a parcel is vested in specific stools, *de facto*, such lands cannot be alienated without the consent of a higher traditional authority (the paramount chief).[6] Customarily, all the land in a traditional area is under the paramount stool. In some areas, such as the Ashanti region, a person cannot process a lease title to land that has been purchased from a community without the express consent of the paramount chief; only with his consent will the Lands Commission even receive such applications.

Both customarily and in practice, paramount chiefs do not have absolute rights over all stool land (see Owusu, 1996; Berry, 2001). Yet paramount chiefs have been recognized by the state as legitimate custodians of stool land and forests. This is because chiefs have legal recognition as 'landowners', though they are required to act as fiduciaries (1992 Constitution). In the management of forestland, even within national forest reserves, the state recognizes chiefs as agents for community representation. In practice, chiefs have acted as both negotiators and signatories to almost all management negotiations, even when they are not explicitly named. For example, a study on community–contractor negotiations of social responsibility agreements in one forest district noted that chiefs dominated the representation of communities, acting as plenipotentiaries in the negotiation, while other non-customary leaders were mere observers (Marfo, 2001, 2004). In practice, chiefs, especially paramount chiefs, have endorsed community consent forms for the granting of timber rights and receive forest revenue (royalties) 'on behalf of communities', as stated in the constitution.

In summary, Ghana has adopted customary law into the formal legal system more so than any of the other cases, to the extent that custom has largely informed the definition of tenure rights and dominates rural land and natural resource allocation. The state has chosen the agency method to recognize customary rights. In practice, the state has empowered chiefs – paramount chiefs in particular – but has failed to enforce the constitutional provision that chiefs act as fiduciaries. The predicted disadvantages of the agency method are apparent, however, and hence the recognition of customary rights has not resulted in an effective integration of formal and customary law.

Burkina Faso

Burkina Faso's natural resources management system reflects the influence of French colonization in its approach to customary law and the civil code. During the colonial period in Africa, France adhered to the principle of transferring lands considered to be 'without owners' to the state. Two kinds of ownership of traditional lands were defined (a titled system for the French and one that recognized remaining customary lands for local people); the powers of traditional authorities were recognized, reinforced and to some extent, coopted. Independence from France was won in 1960. During the revolutionary period beginning in 1983, all customary rights were denied, customary authorities' legitimacy was suppressed so that land could be developed and all private land titles were cancelled. The state was recognized as the only owner of land, including forest.

Since 1991, the need to consider both customary practices and private land titles has been addressed. The 1991 Constitution prescribed respect for customary practices and, in an amendment in 1997, a procedure to harmonize statutory law and customary law. The process of harmonization began in 1996. The decentralization process launched in 1998 has also contributed to this process and the first local governments were established only very recently, in 2006; hence change is still very much in process. The current situation demonstrates that customary systems still have not been integrated with formal law, though there has been some progress.

Tenure rights in statutory and customary laws

Customary and statutory tenure rights have varying degrees of application depending on location. Though classifications overlap, in general, customary tenure rights are dominant and more effective in the rural areas not specifically managed by the state ('non-managed rural areas'). In contrast, in managed lands and urban areas, statutory law is more seriously applied.

The customary rights of communities apply to forested and unforested lands. These lands belong to groups that settled as communities during an earlier period. Prior to colonization, land was the property of different lineages and families. Today, with demographic growth and monetization of land transactions, customary law has evolved to adopt many forms. Generally, tenure arrangements are now shifting from gifts and long-term property loans to rentals and short-term property loans.

Under statutory law, rights depend on administrative title. In rural areas, existing titles include: landownership title, land allocation certificate (*arrêté d'affectation*), occupancy permit (*permis d'occuper*), land-use permit (*permis d'exploiter*) and lease (*bail*).

Reintegration of customary systems of tenure

Burkina Faso's 1991 Constitution grants the state the power to define the process for harmonizing customs with the fundamental principles of the constitution (Art. 101). It also guarantees people the freedom to exercise their customary practices and for this reason several laws recognize customary rights expressly or implicitly. For example, the 1996 Agrarian and Land Recognition Act states that 'the occupation and the exploitation of non managed rural lands in order to provide housing and food needs to the occupant and his/her family, are not subordinated to the possession of an administrative title' (Art. 52).

The state is practicing a kind of essential pragmatism. In the areas where the government is managing land, it tries to maintain control. Where it has no possible control – that is, in the non-managed rural lands – it implicitly allows customary practice to be applied and does not condemn the use of customary rules. This strategy is compatible with poverty reduction objectives. The lack of state intervention also reinforces customary authorities' legitimacy in landownership and management in these areas.

At the same time, the state does not recognize customary landownership in these areas, since the customary landowner now has 'occupier rights' but not ownership rights. Act 031-2003/AN, the mining code, for example, recognizes traditional or customary occupier rights to land when this land is exploited or when the entry on this land causes damage (Art. 65). Traditional pastoralist rights are recognized as well: 'pastoral areas are equally considered as territories reserved for pastures, or *traditional rural spaces* with the object of local operations to preserve or make use of pastoral plans, within the framework of actions pertaining to management of space and of natural resources' (Act 034-2002, Art. 3 [author's emphasis]).

Compatibility in state intervention models

In managed areas, contracts are the principal way that state institutions harmonize statutory and customary laws. In forest management areas, the state uses concession contracts to devolve rights to communities. These concessions benefit village residents who organize into associations or cooperatives, which then have use, withdrawal, management and exclusion rights. Through this devolution, the communities can continue their customary practices as long as they do not contradict statutory law, which means that customary land rights are not recognized. In the forest management areas of Nakambé, for instance, the communities with the concession were applying both statutory law and their customary practices in their efforts to protect non-timber forest products like Shea butter trees. In the classified forest and partial reserve of Comoé-Léraba, the village cluster organized to harmonize statutory rules with customary rights by integrating traditional hunters (*Bozo*) into the management, exploitation and protection of their forest reserve.

In non-managed areas, recognition is mainly through village development councils, which are public entities, unlike the associations and cooperatives formed for concessions.[7] Council members are selected through consensus or election by village inhabitants; two seats are reserved for women. They are under the authority of the local government and are accountable to both the local government and the village population. The council has the authority to allocate land, which allows it to identify customary lands and owners and establish, with them, management rules, with decisions subject to the approval of local government. Formally, then, the council competes with customary authorities. But in practice, customary rules of forest management normally apply.

On the ground, tools used to help promote greater integration of customary rights include the *proces verbal de palabres* and the rural land scheme. The *proces verbal de palabres* helps validate the rights of the owner or user of the land through a public declaration of the rights of customary authorities in the presence of the local administrative authorities. The rural land scheme is a document elaborated through collective research that involves registering, village by village, the customary rights without conflict and on which there is consensus, such that these can then be assimilated into a registry of customary rights. Unfortunately, its use so far has generated conflicts.

In summary, Burkina Faso had a policy of non-integration until village boards were created since 1996. The village boards represented a local land regulation framework, which first became village commissions of non-managed land areas and are now village development councils. The decentralized land board or village council model is used pragmatically by the state in non-managed areas, where customary rights are currently stronger. In forest management areas, the group incorporation model is used to grant concessions to intervillage associations or cooperatives, which can then integrate customary rules for access and withdrawal and forest protection that comply with statutory law.

Indonesia

Dutch hegemony ended after 350 years in 1945. Although Dutch colonialism could be harsh in many forested areas (Peluso, 1990, 1992), the colonial presence was hardly felt in many areas, particularly outside Java, and customary tenure systems functioned with little day-to-day interference. In the postcolonial era, however, the state has shifted from this minimalist model to one of greater intervention, at least where it has the capacity to do so.

Postcolonial phases of statutory law

The Indonesian Constitution states that 'land and water and natural resources wealth are controlled by the state and used for the sake of people's welfare'. In 1960, the Basic Agrarian Law No. 5, developed with Java in mind, recognized traditional tenure systems but required people to register their land – something very few people in the Outer Islands (i.e. outside Java and Bali), where most natural forests are located, were able to do. Many of the law's provisions were also developed, as in many areas of Africa, with permanent, not shifting, cultivation in mind. Indeed, shifting cultivation has been illegal since Dutch times because of its use of fire to clear land. Nevertheless, agrarian law continues to recognize that customary land belongs to customary communities, in contradiction with forestry law (van Noordwijk et al, 2008).

The Basic Forestry Law of 1967 has been more problematic for forest peoples. It stated that 'all forests within the territory of the Republic of Indonesia, including the natural resources they contain, are taken charge of by the State' (Art. 5, Para. 1). During the 1980s and 1990s, under Soeharto's 30-year 'New Order' government, the state classified more than 75 per cent of Indonesian land as state forest, with that figure surpassing 90 per cent for the Outer Islands (Lynch and Harwell, 2002, pxxvii). Soeharto distributed these lands to reward political supporters. Vast areas were allocated first to timber companies, later to industrial timber plantations, followed by transmigration sites, and finally, most recently, oil palm and rubber plantations.

Another national law that interfered significantly with customary management of forests was the Village Governance Law, which mandated at least superficial adherence to a standardized form of local governance across the nation, thus undermining the authority of customary leadership. Bennett (2002, p60) describes this law as intended to 'subvert traditional forms of governance'.

In 1999, the Basic Forestry Law was revised, with more provision for local management. The existence of customary communities, cultures and forests, for instance, was recognized. Communities were granted the rights to help determine the size of their forest area, collaborate in monitoring, be protected by the government from pollution and deforestation caused by others, and more. Wollenberg and Kartodihardjo (2002, p88) note, 'Scattered throughout the law are references that suggest that forests should be managed according to principles of social equity, empowerment of customary communities, fairness, property, and sustainability.' But these authors also see 'escape clauses' that leave ultimate power with the state. Van Noordwijk et al (2008) note that no communities have yet managed to obtain formal recognition of their customary forests.

Also in 1999, the country embarked on a decentralization process (Barr et al, 2006). A new law (UU No. 22) delegated governance authority to autonomous regions (provinces, districts and municipalities) and granted districts and municipalities authority and responsibilities that explicitly included agriculture, environment and land. But the following year, Regulation No. 25/2000 defined the mechanism by which the central government could resume authority when autonomous regions were deemed incapable of carrying out their tasks, thereby reaffirming the Ministry of Forestry's dominant role in forestry policy and planning (McCarthy et al, 2006).

Over the past decade, there has been an ongoing tug of war between the central government and the districts on forest management authority. Different districts have opted for different strategies, both in their interactions with the centre and in their chosen trajectory for the future. Some districts in Jambi, for instance, have taken a conciliatory attitude towards renewed state control; those in West Sumatra have been more intransigent about keeping the rights they gained in the original decentralization law. Districts in both provinces have opted to rejuvenate their customary systems (*nagari* in West Sumatra, Raharjo et al, 2004; and *rio* in Jambi, Hasan et al, 2008).

Diversity of customary law: Two cases

Indonesia, an archipelago of some 17,000 islands spread over 1.9 million km^2, is probably the most ethnically diverse country in the world, with some 742 languages[8] and 283 million people representing more than 300 ethnic groups.[9] The customary tenure systems in forested areas are correspondingly varied. We have chosen to describe two briefly.

The first is the Uma' Jalan Kenyah of Long Segar, a group of *dayaks* (indigenous people of Borneo) living in the centre of East Kalimantan. They are bilateral swidden agriculturalists,[10] originating in Long Ampung near the Malaysian border. When Long Ampung's population grew to about 1000, the community split and some members headed for a new area (consistent with general Bornean custom in this sparsely populated area). In Long Segar, the migrants discussed land availability with the few people already there and determined an area – marked by rivers – that all agreed the migrants could claim as their community territory.

Each family then began clearing primary forest for rice fields and that land became theirs to use and pass on to their descendants. Their rights to the land are not, however, absolute. When a family moves away and there is no direct descendant of the original migrant in the community, the rights return to the community for distribution. Community members can also claim rights to individual plants and trees in unclaimed community forest by marking them as their own; anyone who plants something has the right to harvest it, even if it is on someone else's land, and people who pass by and feel hungry have the right to harvest anyone's plants to assuage their hunger. Men have a tradition of expedition making (*tai selai*) and are often away from home, leaving women to manage the farm (Colfer, 1985a, 1985b).

Community leaders have the right to make agreements with outsiders about land use, though they usually discuss such arrangements with community elders and also with the community at large, resulting in consensual decisions. One source of persistent conflict within Kenyah communities is suspicion of leaders and their ability to capture illicit rents from such interactions – which often happens. Notably, the state has subverted customary rights by granting two timber concessions and promoting a resettlement scheme on Kenyah lands over the past several years.

In rural West Sumatra, among the Minangkabau living in the 'frontier' (*rantau*) with the neighbouring province of Jambi, there is a quite different system.[11] The Minangkabau of Pulai are organized into three matrilineal clans (Ghana's Ashanti represent another matrilineal group). Again, newcomer clans had to ask permission from the long-settled clan members to settle there and the newcomers were allocated certain lands. The local system depended on paddy rice, which was owned and transmitted via the matrilineal clans, with control primarily in the hands of the men (brothers) in that clan, though women were the primary rice cultivators. These lands were inherited by the clan sisters' children. Upland fields, which belonged to the men who cut them from the forests, were typically owned by the nuclear families of those men and passed on to their own children. These upland fields were first planted with rice and then with rubber during the 1980s. Such upland fields, unlike the clan lands, could also be alienated by the men who cleared them, via sale or rent. Sharecropping arrangements for rubber tapping (with a third of the harvest typically going to the owner) were also common in the 1980s. In response to market demand, this rubber was later converted to vast fields of oil palm; the paddy rice remains and those who can are returning to their swiddens now that the oil palm price has fallen drastically (late 2008).

Alienating clan lands required agreement of every member of the clan, something almost impossible to obtain, since Minang men, like Uma' Jalan men, are famous for making expeditions (*merantau*) and are often away. Clan lands have, however, routinely been alienated by the central government, which in this area established a transmigration site of several hundred thousand ha, beginning in the late 1970s, which completely surrounds Pulai; Kerinci Seblat National Park was established on lands belonging to related clans.

Though both of these once-isolated communities have maintained many of their customary practices, in general Indonesia is moving away from Fitzpatrick's minimalist approach. There are conflicting pressures, respectively political and economic, towards the agency model and the group incorporation model. Government personnel tend to deal with the formal leader of villages in a manner that is consistent with the agency model, while mechanisms exist for formalizing community groups into corporate entities (cooperatives) specifically for interaction with industry, such as logging and plantation companies. Overall, statutory law, which previously denied all customary rights in forestry, has opened some tentative opportunities for legal integration, but in general the Indonesian state appears to have little interest in pursuing this option.

Guatemalan Highlands

In Guatemala, the state's commitment to recognizing customary law or finding effective ways to integrate customary and formal systems appears relatively limited. Highland communal forests in particular have been highly vulnerable to external threats, including threats from the state itself – unless communities have been able to obtain formal land titles.

Social construction of tenure rights

Rights to tenure and natural resources in Guatemala have been shaped by five historical developments. First, community rights arise from the permanence of indigenous peoples on the lands that they occupy today and manage through deeply rooted customary norms. Second, colonial agrarian policies reorganized tenure rights by usurping lands from original peoples, generating administrative disorder; this continues to be the primary cause of agrarian conflict in the country, even though colonialism ended in 1821.

Third was the creation, at the turn of the 20th century, of modern institutions intended to secure tenure rights, above all for large property holders, and resolve the administrative disorder. This involved the creation of a property registry and the Civil Code, through which the state obtained greater fiscal control over property but at the same time legitimated and facilitated the seizure and usurpation of indigenous peoples' communal lands. The fourth development involves land struggles, which began with a failed agrarian reform attempt in 1952 and continued with colonization programmes in the northern lowlands after 1960, the evacuation and following resettlement of the population affected by armed conflict and the creation of land access programmes. The fifth development is the territorial conflicts of the past 15 years, related to the state's redefinition of tenure rights in favour of mining, energy, agricultural and conservation projects.

Today, tenure rights have distinct meanings depending on whether they are private, communal or state. On the one hand, private property is guaranteed by Article 39 of the constitution as an inherent human right that the state is obliged to recognize. Under this mandate the idea that private property is absolute has spread. Communal lands, on the other hand – though they have special state protection and the possession of communal lands is guaranteed by the constitution – are, in practice, legally precarious. This is because communal lands have no specific normative framework, even though this was also mandated by the constitution. In contrast with the discourse surrounding private property, the idea that communal lands and forests are open access resources is widespread in Guatemalan society. Public policies privilege private property and exert pressure for the dissolution and transformation of communal tenure. Nevertheless, in the highlands, communal tenure is deeply rooted in several land types: communal lands, municipal lands, cooperatives and *parcialidades* (a form of communal tenure defined by kinship).

Communal forests, in particular, may be formally owned by communities but are more likely to be state lands, under municipal governments. State lands are those for which rights are the least defined and delimited, since they have traditionally been considered open access resources, or because they overlap with forms of communal tenure.

Making rights matter: Formal and customary norms

The bundle of rights for access, use, management, exclusion and alienation of lands and forests has been constructed through two parallel principles. On the one hand, large landholders and those with means mobilize the legal and institutional apparatus of the state for the exercise and formal recognition of rights. On the other hand, customary mechanisms are still in force for the exercise of tenure rights at the local level, whereby communities organize individually and collectively, as well as distribute harvest quotas and responsibilities among their members.

Communal tenure includes collective rights that correspond to the community as a whole, such as access to forests, water sources, sacred places and pasture. The rules for access and harvest are distributed exclusively among group members, who are easily recognized by their sense of belonging, through mechanisms based on collective responsibility and participation. For example, to have rights to obtain forest products, individuals must have contributed to the tasks and duties of their community. Individual rights to communal lands are assigned and recognized according to local norms.

From the point of view of the state, rights can be exercised and claimed only according to formal laws. In practice, these two mechanisms for the exercise of rights are not isolated but rather are closely related – not because of state recognition policies but because of community adaptations. For example, communities that have registered their properties have a greater possibility of exercising their customary rights than those that have not. Similarly, community organizations that have formalized their existence, as subjects of rights by obtaining legal standing (*personería jurídica*), have a greater possibility of exercising their collective rights.

Struggle for recognition of customary rights

In 1996, in peace accords that put an end to more than three decades of armed conflict, the state promised to recognize customary rights. To this end, the Municipal Code was reformed and laws for participation, decentralization and development councils were passed. The Accord on Identity and Rights of Indigenous Peoples recognized the participation of community authorities and extended the application of customary norms to rights relating to land and conflict resolution.

Despite some progress in the application of indigenous rights, the state continues to promote or tolerate actions that infringe on community tenure rights, such as the authorization of mining licences in indigenous territories and the failure to recognize the binding nature of community consultations. In 2007, a proposal by environmental groups to create a water law was rejected by highlands indigenous communities for failing to recognize existing customary norms.

Though forest management is governed by the Forestry Law, in practice communities have established their own norms. Conflicts have arisen between community authorities and the Institute of Forests in this regard. These are mainly related to requirements for licences for resource use, particularly timber and firewood. The forest agency has insisted that communities obtain permits in accordance with formal regulations and has facilitated the creation of municipal forestry offices to make this easier; it considers all other forest use illegal. Conflicts arise because not all forest users obtain these licences, mainly because they are low-volume harvesters and are often authorized or tolerated by community authorities. In other cases, the municipal forest office grants licences to communal forests without the consent of community authorities. Some communal forests have been converted into protected areas, where customary norms have been replaced by official conservation norms and management plans. The resulting limitations on traditional use hurt the poorest families, who depend most on forest resources.

Overall, the Guatemalan state has done little to recognize customary rights despite the constitutional provision protecting communal lands; hence it is not possible to identify a model of recognition. For their part, and partly to defend their rights, some highlands communities have sought to incorporate as formal organizations or seek other ways to register their properties formally. This has allowed them greater margin for traditional practices, though forest management remains highly regulated.

Both communities and some state entities, however, are making efforts to obtain greater complementarity between formal and customary norms. For example, several communities have obtained access to forest incentives and created communal forestry offices. Progress has also been made regarding the formulation of a strategy for conservation and management in communal lands, which proposes the recognition of individual and collective rights and local systems of organization and governance. Another proposal includes the reformulation of current categories of protected areas to make them more relevant to communal lands and the rights of indigenous peoples.

Challenge of integration: Analysis of cases

All the cases have demonstrated that the boundary of what has been called customary land law is fluid and evolving from what might be termed its historical traditional practice. And each shows some level of legal integration in the governance of tenure, but with wide variation in models and interests of the state.

The Ghana case most clearly represents Fitzpatrick's agency model, with substantial state recognition of and engagement with community representatives, in this case chiefs. Overall, efforts appear aimed at accommodating customary law, rather than ignoring or subverting it. The challenge for policy innovation in realizing the constitutional provisions of this recognition is how to ensure accountability by 'compelling' these representatives to act indeed as fiduciaries.

Burkina Faso represents the group incorporation model in state managed forests but with the land board or village council model outside formal management areas. After a period characterized by the suppression of rights under the revolutionary government, the constitution today calls for harmonization, and certainly some efforts appear to have been made in that regard. In general, the state still demonstrates ambivalence towards, and perhaps some ongoing attempts to suppress, customary land rights (at least until recently) but is receptive to accommodating local forest practices.

Indonesia represents a complex mix of minimalist, agency and group incorporation models. In part because of the nature of the country and the population, the situation is confusing: the decentralization process has given rise to a multiplicity of approaches. This is also the case in Cameroon (Diaw et al, 2008; Oyono et al, 2008). In general, however, as in Guatemala, the state shows little interest in accommodating customary law and land or forest rights, with a tendency towards suppression to the extent that the state is present in a particular region.

Guatemala has not, in practice, recognized customary rights to land and forests and demonstrates efforts at accommodation or integration only on a small scale or through specific state entities or actors in some communities. It remains to be seen whether the pending national proposals will progress beyond the idea stage. For their part, communities have sought to defend their rights by adopting formal institutions, through group incorporation for example, and by obtaining land titles.

Although Fitzpatrick emphasizes the importance of the source of land tenure insecurity in defining the model, this aspect does not appear to have played any role in the choices made in each country. Rather, they appear to have been made by default, or perhaps to meet the state's needs. For example, the minimalist approach is used in Indonesia – as we suspect in many other countries – only where the state is unable to enforce its will, rather than because this is the best way to secure tenure rights for these communities. The agency method almost by definition is primarily advantageous to the state rather than the community, at least without mechanisms for downwards accountability. Other options appear to have arisen, as in Guatemala, mainly as a defence mechanism, sometimes against the actions of the state itself.

The group incorporation method appears particularly relevant for outside contracts, especially for logging and other forest management activities, at least in Indonesia and Guatemala. In Burkina Faso, however, the group incorporation model is used to give formal concessions to community organizations for forests they have used traditionally – without recognizing land rights; in contrast, the land board or village council model is used where the central government has less reach, and here the model has more potential for integrating customary land rights.

In summary then, the choice of model – and, therefore, more or less state intervention – does little to suggest greater or lesser commitment to supporting and integrating customary rights. That is, the model of recognition itself does not suggest whether the state's overall approach is ambivalence, accommodation or suppression. On the other hand, the two countries with the most suppressive policies, Indonesia and Guatemala, have been the most effective at avoiding any commitment to a particular model of recognition. In the former, multiple models abound in different spheres, as they are found convenient or expedient for the issue at hand; in the latter, communities have largely adapted to statutory law to find ways to maintain customary rights and practices.

One aspect that may provide greater insight into governments' commitment to customary forest rights is the question of exclusion. Only in Ghana do communities have *de jure* exclusion rights, through their customary authorities (in Burkina Faso, customary authorities have *de facto* exclusion rights, and even the state generally asks for their consent before making forest-related decisions). The granting of exclusion rights may not always be straightforward and can be difficult to address in situations of multiple users or overlapping claims; for example, granting one community exclusion rights often results in the exclusion of other customary users, especially temporary or seasonal users such as pastoralists (see Chapter 9). Exclusion also raises issues about authority and representation (see Chapter 5). Nevertheless, the right to exclude logging companies or concessions that enter forests with state authorization would appear essential. In Burkina Faso, Indonesia and Guatemala, the consent of the community is not required for the state to authorize licences to others in these same forests. This makes it much more difficult for communities to protect their forests and any effective forest tenure reform to benefit communities should, at a minimum, require the state to obtain the prior consent of the community before issuing permits.

The cases also provide a glimpse of the complexity of integrating statutory and customary systems. Customary tenure rights to, and practices in, forests are embedded in rights to land and shaped by social and cultural relations defining group membership (among other things) and governed by customary authorities. Different ethnic groups and communities give rise to different sets of rights and practices, hence in countries with greater social diversity, like Indonesia, customary laws are also very diverse. The five spheres constituting the bundle of rights provide additional dimensions upon which rights can be allocated – whether shared or divided. Statutory systems are only somewhat less complex, at times (or even frequently) promoting contradictory policies, as in the case of the recognition of customary rights in Indonesia's agrarian versus forest sectors.

In all the cases we find some level of statutory recognition of customary rights and in some cases the substantive practices of tenure rights have been informed by customary law. For example, in Ghana, all the rights that can exist in land as recognized by the state had previously been customary laws. Today, customary rights are invoked even in the courts in addressing land-related disputes. In Burkina Faso and Indonesia, even though the state has *de facto* ownership of land, the daily uses to which forests are put have relied largely on customary practices – except in the increasing number of cases in Indonesia where large concessions are given to private enterprises, which then have the right to usurp local lands and forests. Even in Guatemala, where indigenous communal lands have historically been usurped by the state and private interests, and state forestry law has criminalized many local practices, creative innovations continue to arise to accommodate customary practices.

In their critical review, Kasanga and Kotey (2001) endorsed the pluralistic path of the tenure systems existing in Ghana observing, for instance, that completely overturning the customary system is impractical and unworkable. The other cases appear to support this view, given the simple staying power of customary practice, the revitalization of customary systems under decentralization and the common *de facto*, if not always *de jure*, recognition of local land management.

Thus the argument that policy innovations should focus on legal integration, to ensure the positive elements of both systems are married, is well supported here. In theory, legal integration would exist when the elements of the two systems either reinforce each other or perform complementary functions valued by local people. The interest in integrating formal and customary forest tenure systems derives from the difficulties that plague forest-dwelling people who have had to struggle with the incompatibilities between their own customs and the requirements of their formal governments. In this struggle, they have often been the losers.

Despite its theoretical appeal, however, how to achieve integration in policy and practice is unclear. A particular challenge is how to give material meaning to the form that the legal integration of forest tenure will take in practice and the extent to which it can be empirically delineated to stand on its own as a model. The four models of recognition examined here remain inadequate.

One point of departure is when statutory recognition has moved from discourse to practice and is clearly aimed at accommodation and integration, rather than suppression. The lack of political will to pursue integration can be a major obstacle and the lesson of the highlands in Guatemala is a good example. Given the power of the state, it remains to be seen how an integrated system can be built to protect local rights, particularly over high-value resources. The cases suggest that a state that leans towards suppressing customary systems, as in Indonesia and Guatemala, could adapt a model of recognition or integration to promote this goal instead.

At the same time, the lack of political will for integration may come from customary actors themselves, especially if integration will shake the power and economic interests of local elites. For example, pursuing full recognition of customary land law in Ghana may necessitate new procedures for benefit sharing of forest revenue and require chiefs to act as fiduciaries and not private landlords. This raises issues regarding the kind and degree of state intervention that is acceptable or appropriate when customary practices are inequitable, discriminatory or undemocratic – for example, when women's right to inheritance and downwards accountability are not built into traditional structures or practices.

Other challenges include the very complexity of customary systems of practice themselves. Most customary laws are not codified, leading to a large plurality of interpretations and applications. The extent to which the state can recognize customary laws and integrate them will depend on how easily such rules are organized. Though some call for codification of these laws, others argue that this would interfere with the essential fluidity and adaptability that characterizes customary law over time.

Finally, in this chapter, we have dealt mainly with communities whose members share a customary system. In fact, residential groups or villages are increasingly likely to comprise several ethnic groups or 'communities', each with its own customary rules. The interests of minorities must be taken into account in crafting any integration of customary and formal forest tenure systems. Stakeholder negotiation becomes particularly important in such contexts.

Notes

1. Central concepts are defined in Chapter 1.
2. Indeed, the state itself often includes self-contradictory components.
3. At the same time, we see a potential land mine in her call for 'designated spheres' and her emphasis on 'private property', in light of the socially embedded nature of customary systems as summarized above.
4. Elbow et al (1998), cited in Diaw (2005, p51), uses a different framework for organizing policies in 22 African countries: (1) non-recognition or abolition; (2) neutral recognition; (3) recognition aimed at replacement; and (4) zoning recognition.
5. 'Stool' refers to the seat of a chief of an indigenous state and represents the source of his (rarely her) authority.
6. Traditional areas consist of several communities, which may have their own stool lands and chiefs. The various community chiefs in the traditional area, in addition to the paramount chief, constitute the traditional council, which is presided over by the paramount chief. Chiefs are male but are accompanied by a female queen mother, who in rare cases may occupy the stool and thus act as chief.
7. Village development councils may also play a role in managed forests that are not classified forests and are hence under communal rather than state domain. If the managed area is under concession, the council's role is more limited, but if it is not under concession it will be assigned to the council.
8. The determination of what is a language and what is a dialect remains an issue in Indonesia (www.ethnologue.com/show_country.asp?name=id) (last accessed September 2009)
9. See http://en.wikipedia.org/wiki/Demographics_of_Indonesia (last accessed May 2009)
10. In bilateral peoples, descent is traced through both the father and the mother.
11. See Colfer (1991) and Colfer et al (1988, 1989) for this particular setting.

Part III
Governance Institutions: Authority Relations and Social Movements

5

Authority Relations under New Forest Tenure Arrangements

Anne M. Larson, Emmanuel Marfo, Peter Cronkleton and Juan M. Pulhin

The question of authority appears to be a central factor affecting the outcomes and success of forest tenure reform, yet this issue has not been well developed in the related literature. For communal properties in particular, decisions regarding 'authority' are central to shaping how decisions are made, whose opinion or knowledge is taken into account and how access to land and natural resources is determined in practice. When property rights are formalized, authority relations define the extent of decision-making power that is held at different levels, from the community to the state, and the way in which customary and *de facto* local management norms and knowledge regarding resource management are – or are not – recognized in the formalization of tenure rights and institutions. Authority relations are also important in understanding on-the-ground dynamics of power, which shape access to resources and benefits.

The term 'authority' is used in several ways, particularly in the realms of policy and practice. In particular, it is used to refer both to the abstract notion of power and to the person or institution holding that power (Fay, 2008). According to Weber (1968), authority refers to power that is 'legitimate'. The issue of legitimacy raises additional questions, of course: who considers the authority legitimate, for example, and what constitutes legitimacy? This chapter considers authority not as a fixed attribute that can be mandated or assumed, but rather as something that is constructed through social interactions and subject to conflict and contestation (Sikor and Lund, 2009). It argues that the recognition of community rights by central governments leads to political contestations over authority as social actors react to the changes introduced by the reforms. The central issue of concern here is the institution selected to represent the collective that receives formal rights under these new legal arrangements.

Both the nature of the institution representing the collective and its domain of powers are fundamental to the distribution of access to land and forest resources and to the benefits they generate. The institution chosen to represent the collective by law may or may not be considered a legitimate, representative leader by the population, and it may or may not be the same one that has played this role or made these decisions in the past. This institution may be bestowed with the power to make significant external and/or internal decisions on behalf of the collective regarding resource access. It may be in charge of resources, including financial resources, intended to benefit the collective.

This chapter explores these issues specifically with regard to indigenous territories in Nicaragua and Bolivia, ancestral domain lands in the Philippines and local forests in kin-based stools in Ghana. All of these cases involve the recognition of customary rights to land or forest resources and the empowerment of traditional local actors and the institutions that represent them. In these four cases, the politics of authority takes different forms. Together, they demonstrate that recognizing tenure rights is not a straightforward process that simply grants greater legal security to a set of existing, and fair, customary institutions. Rather, recognition alters the existing institutional structure through the act and practice of recognition (Ribot et al, 2008); it is therefore a highly political act that is subject to negotiation, contestation and manipulation. The cases demonstrate that these contestations may lead to conflict over authority or to new, emergent configurations that improve local resource governance.

The cases presented here demonstrate:

1 the crucial role of the institution established to represent the community's formal tenure right in shaping the exercise and allocation of rights and benefit distribution in practice;
2 ways in which the process of recognizing rights through specific institutions turns those institutions into sites of struggle such that they may break down or be manipulated under the pressure of competing interest groups; and
3 the central role of representation and downwards accountability in these processes.

The chapter draws on the theoretical development of 'authority relations' in Fay (2008) to argue that the construction of the property right in practice is about much more than just choosing the correct, downwardly accountable institution to represent the population. Rather, the construction of authority – and sometimes multiple authorities – emerges from a process of contestation involving a dynamic tension among the state, the community and the entity chosen by each of these to 'represent' the community.

Authority relations and communal forests

The subject of authority has been identified as an important emerging issue with regard to property research and practice and is integral to understanding the interplay of tenure and rights in forest tenure reform. As Ribot et al (2008) argue, 'while "property" is an enforceable claim (MacPherson, 1978), too much attention is trained on the rules of the game rather than the origins and construction of the authorities "enforcing" the rules'. If property refers to the rules of the game, then the implementation of new tenure rights refers to the making of those rules. The authority relations established in that process define the extent of decision-making power that is held at different levels and by different institutions and shape access to resources and benefit distribution on the ground (Larson, 2008; Larson et al, 2008; Sikor and Thanh, 2007).

The recognition of tenure rights for communities, then, results in numerous sites of struggle. This chapter focuses specifically on the choice of institution to represent the collective. When those receiving new or formal rights already have customary rights to the land, it might seem that the simplest solution is to recognize the institution that is currently in power. However, there are at least two problems with this.

First is the question of tradition and the issue of 'traditional authority' in particular. The call to respect customary rights, such as traditional land rights, has been central to indigenous struggles in Latin America. But tradition and custom are loaded terms. For some, respecting or recognizing tradition refers to the enfranchisement of peoples whose rights have been denied (Taylor, 1994); for others it means the opposite, protecting people as a group but not individual rights – a necessary condition for citizenship (Mamdani, 1996; see also Ribot et al, 2008).

Ribot et al (2008) warn, in particular, against conflating customary rights or practices with customary authority.[1] When the state recognizes, in the tenure reform, a particular institution as the community representative, it is granting that institution external legitimacy. This institution may not have internal legitimacy, or it may have internal legitimacy but not to manage the particular set of powers now being granted (Fay, 2008). An example is the recognition of non-democratic institutions – chiefs and headmen who inherit their posts – in African nations undergoing decentralization (Ribot et al, 2008; Ntsebeza, 2005), and this issue is relevant to the Ghana case here.

Second, the granting of tenure rights may necessarily involve the formation of new institutions, particularly for large territories. Indigenous movements in several Latin American countries, including Bolivia and Nicaragua, have promoted a territory model comprising multiple communities for the implementation of indigenous property rights (see Chapter 2). These territories are expected to facilitate the demarcation and titling of large areas covering the land areas that indigenous peoples have used historically. The territory model – seen as the most advanced form of granting indigenous tenure rights – should permit sufficient space for resource conservation, use and management, real participation of indigenous peoples in the definition and demarcation process and the use of resource management models that combine traditional and modern practices for long-term development (Davis and Wali, 1994). The Philippines case also represents a territory, known as ancestral domain, comprising multiple communities, though the total area is much smaller.

The territory model has encountered serious problems, however, due to the choice of institution to represent the collective. When such territories are newly created, their demarcation and titling require the formation of new governance institutions. In his review of experiences in the legalization of indigenous territories in four South American countries, Stocks (2005, p98) argues that 'the weakness of the indigenous governing institutions', and particularly the lack of democratic representation at the territorial scale, 'is an extremely vulnerable aspect of the indigenous land movement'. This problem is explored in Bolivia, Nicaragua and the Philippines.

Fitzpatrick (2005) highlights four ways in which the state can recognize customary tenure (see Chapter 4), each suggesting a specific institutional model for representing the collective, with increasing levels of state intervention in existing arrangements. The two that are relevant for the cases here are the agency and the group incorporation approaches. In the former, an actor – commonly a chief or clan leader – is selected to represent and negotiate in the name of the collective. The latter involves the formal incorporation of the group into a legal entity, with corporate bylaws and formal internal rules of practice. In the cases studied here, Ghana represents the agency approach and Bolivia and the Philippines, the formal incorporation approach. The Nicaraguan case falls somewhere between the two.

As the case studies will demonstrate, the implications of these models of recognition for the construction of authority are by no means predictable but are, rather, the result of conflict. Fay (2008) develops the concept of *authority relations* based on three sets of interactions: between an actor in authority and the subjects, between this actor and the external justification for his or her rule or power, and between the subjects and this external justification. If we assume that the 'external justification' generally refers to the state (and statutory law), we can use this concept to refer to:

1 the relationship between the community and the state;
2 internal relations between community members and the community actor(s) or institution in authority; and
3 the relation between this institution and the state.

These arenas of social interaction are useful for analysing the case studies to understand the roots of conflict and the construction of authority as legitimate power.

The first set of relations plays out through the central role of the state in establishing and implementing the legal framework granting local tenure rights. This relationship between the state and the community then leads to the question of who receives powers when the state recognizes or transfers tenure rights to communities, and the implications of that choice. Hence the other two sets of relations immediately raise an important question regarding the community institution in authority, since this is a major site of contestation. For example, the state and the community may recognize *different* actors as the 'legitimate' representative of the community, as in the Nicaragua and Bolivia cases below. The next section presents the four cases and the following section returns to a discussion of the models adopted and the ways in which these three sets of interactions help us understand authority relations as a site of struggle affecting outcomes of reforms.

Four case studies

The cases describe a variety of situations in which tenure rights are granted or formalized. In each, a particular institution is recognized as the formal representative of the community. The process of implementation demonstrates how this institution shapes the distribution of rights on the ground and becomes a site of contestation.

Nicaragua: Indigenous territories[2]

Nicaragua's 1987 Constitution recognizes and guarantees the rights of indigenous and ethnic communities to their cultural identity, forms of organization and property, as well as to the enjoyment of their waters and forests. The Autonomy Statute (Law 28), also passed in 1987, created the North and South Atlantic autonomous regions, whose first regional autonomous councils were elected in 1990. These two regions represent approximately 45 per cent of the country's land area and only 12 per cent of the total Nicaraguan population but are home to the vast majority of indigenous and ethnic people, who constitute 8.6 per cent of the nation's total (INEC, 2005). According to the 2005 national census, the Miskitu population is the largest, with 121,000 people, followed by the Creoles, with 20,000 and the Sumu-Mayangna with 10,000.

In 2003, the Communal Lands Law[3] (Law 445) was enacted. Like the constitution, this law formally recognizes the rights of indigenous and ethnic communities to their historical territories, but it also establishes the institutional framework for demarcation and titling and for the formal recognition of indigenous leadership institutions ('communal authorities'). The law responded to demands of Caribbean Coast[4] indigenous communities and, more specifically, commitments acquired by the government of Nicaragua in a ruling by the Inter-American Court for Human Rights in *Awas Tingni v. Nicaragua*. The Sumu-Mayangna community of Awas Tingni filed the suit against the government for granting a forest concession, on their traditional lands and without community consent, to the Korean company SOLCARSA in 1995. The community's legal representatives had fought the concession in the national courts to no avail, despite a Supreme Court ruling in 1997 that the concession was unconstitutional for failing to obtain the prior approval of the regional council, as established by law (Wiggins, 2002).

In 2001, the international court ruled in favour of Awas Tingni, finding that the Nicaraguan government had violated the American Convention on Human Rights as well as the community's rights to communal property as guaranteed by the Nicaraguan Constitution. The court ordered the state to create an effective mechanism for demarcation and titling for indigenous communities 'in accordance with their customary laws, values, customs and mores' (judgment cited in Anaya and Grossman, 2002).

The Communal Lands Law guarantees indigenous communities 'full recognition of rights over communal property, [and] use, administration and management of traditional lands and their natural resources' (Article 2). Nevertheless, demarcation and titling proceeded extremely slowly from 2003 to 2006, with little enthusiasm from the central government and accusations of corruption in the intergovernmental institution established to oversee the process. The process picked up again after January 2007 when the Sandinista political party (FSLN) returned to power.

The law establishes procedures for titling either as a single community or as a group of communities. It formally recognizes 'traditional communal authorities' as the legal representative (externally) and government (internally) of the community (Art. 3). These include the *wihta* (communal judge), *síndico* (the official most often in charge of land and natural resource allocation today), and others. When communities form multicommunity territories, the territorial authority is elected by an assembly of all the communal authorities from participating communities, according to the procedures they adopt (Art. 3, 4). This new governance institution is the administrative organ and legal representative of the territorial unit (Art. 5). The regional council then registers and certifies the people elected. The elected community-scale institution authorizes the use of communal land and resources by third parties; the territorial-scale institution authorizes the use of resources common to the multiple communities of a territory (Art. 10).

Some Caribbean Coast indigenous communities have a century-old history of close association as a group of communities; other affiliations are more recent and based on common history and/or social and economic relations. For example, younger communities were sometimes formed by family members from an older 'mother community' to increase access to forest lands and resources. When the Caribbean and Central American Research Council (CCARC) conducted a participatory study to identify indigenous territorial demands in the late 1990s, the formation of multicommunity territories was recognized as a new priority in the conception of land rights and autonomy (CCARC, 2000), though this vision was still incipient (Hale, personal communication).

Today, this vision has become the norm. At the time of the study, one of the two sites studied in the CIFOR-RRI research project, Tasba Raya, which has seven recognized Miskitu communities, preferred to be demarcated as individual communities but later, in 2005, formed a territory and a territorial authority. The other site, Layasiksa, which actually constitutes two Miskitu communities, preferred to be demarcated as the community of Layasiksa when our study began; in January 2009, however, Layasiksa joined with a third community to form the territory Prinzu Rau. Both groups decided that forming a territory was in their strategic interest: Tasba Raya communities joined together to face a conflict over territory with a neighbour, Awas Tingni, and Tasba Raya and Layasiksa formed territories to gain a greater voice in negotiations with regional political leaders.

For their part, political leaders from the Miskitu party Yatama have been pushing communities to form territories based on a design of their own conception. According to Miskitu leaders, they are interested in forming territories that cover a significant part of the land area, leaving little behind as 'national land', including all indigenous communities inside territories and moving quickly while the political moment is favourable (CRAAN, 2007). Yatama is also interested in reshaping electoral districts; this involves eliminating the municipal structure imposed by the central government and replacing it with an 'indigenous' structure of territories and territorial authorities. This would involve legal reforms that are currently in draft form.

In theory, if community self-government were the foundation, with multi-community territorial institutions at the second tier and electoral districts based on these structures for the election of the regional autonomous councils, this new governance structure could provide the institutional basis for the self-determination of the indigenous and ethnic populations of the autonomous regions. But not all indigenous and ethnic groups, even many Miskitu, feel represented by Yatama or trust its leaders' motivations. What has happened in practice demonstrates the intimate relationship between the issue of territory, the institution representing that territory and access to resources.

Síndicos, the community actors in charge of land and natural resources, have been notoriously corrupt. All over the autonomous region, they have been accused of selling land to colonists, selling timber to loggers and intermediaries and failing to account for these deals to community members. In addition to representing their communities or territories in outside negotiations, *síndicos* also have access to a percentage of tax funds from natural resource exploitation that are designated for the communities from which resources were extracted. Their only punishment for corrupt practices has been removal or not being reelected to their post. In Layasiksa, a *síndico* absconded with part of the tax funds after he had been replaced.

Tasba Raya and Layasiksa had both worked diligently over the recent years preceding this study, in part with external support, to elect more responsible leaders and improve local norms, sanctions and accountability systems. The *síndicos* in both territories have been reelected and are free of such accusations. But both territories have had direct conflicts with regional government officials. Since the election of its territorial authorities, Tasba Raya has been unable to obtain registration by the regional government. The lack of accreditation means that it cannot undertake any external activities in legal representation of the territory, including, for example, having access to the tax funds.

Though he had been previously certified as *síndico* of Layasiksa, the *síndico* of what is now the Prinzu Rau territory has had similar problems, because reaccreditation occurs every year. It took most of 2008 to gain accreditation and he was certified for only five months. In both cases, regional government officials rejected the communities' petitions, stating that the 'territorial authority' they were proposing did not conform to the model established in another territory – even though the law states that the communities in each territory should determine the nature of their own authority.

That may not be the real reason for rejection, however. Based on the vision of Yatama leaders, both Tasba Raya and Layasiksa–Prinzu Rau are subsumed into much larger territories under the administration of territorial coordinators, apparently selected by the head of Yatama. Though the law mandates that 'territorial authorities' be elected in territorial assemblies, elections that had taken place in some territories were evidently manipulated. In one case the people certified were not the ones elected. In other cases, the territory coordinator is unknown to community leaders, who clearly did not participate in his election.

The territorial authorities elected by the people of Tasba Raya and Prinzu Rau have had trouble getting certified because they interfere with political leaders' larger project. Even with certification, the Layasiksa–Prinzu Rau *síndico* has never had access to the tax funds, since this power has been assigned to the coordinator of the larger territory established by regional leaders. Though this *síndico* has maintained his role in approving permits for logging, which has expanded substantially for salvage operations in the wake of Hurricane Felix (September 2007), in many territories this power is also now in the hands of the territorial coordinator. If Yatama's vision of indigenous territories does result in the remaking of the municipal administrative structure, then the nature of the territory and the institution representing that territory has further political and economic consequences for the distribution of community resources.

Bolivia: Guarayos community land[5]

The recognition of indigenous land claims in Bolivia has resulted from a slow process of policy reform driven by rural collective action to pressure government decision-makers. Indigenous people developed an activist social network of organizations in the country's eastern lowlands under an umbrella organization, the Confederation of Indigenous People of Eastern Bolivia, and used mass marches and other forms of protest to draw attention to their cause. This pressure led to constitutional change, the signing of the International Labour Organization's Convention 169 and a series of presidential decrees defining a type of indigenous property known as original community land (*tierra comunitaria de origen,* TCO).

Demarcation of the TCO in Guarayos has taken more than a decade. Guarayos is a rapidly changing forest frontier province in the north of Bolivia's Santa Cruz department, with an interdepartmental highway that opened the region to outsiders, including timber industries, ranchers, large-scale commercial farmers and smallholder colonists. Many of these actors, who became competing stakeholders for land, were moving to the region for its fertile soils and expanses of forest. By the 1990s there was growing tension in the province where indigenous people began to feel the pressure as others claimed land and extracted resources (for a full description of the Guarayos TCO, see Chapter 3).

The Guarayos people are represented by the Central Organization of Native Guarayos Peoples (COPNAG), created in 1992 to pressure the government to recognize their land claims. COPNAG overlies two grassroots organizational structures: one a remnant of institutions imposed by religious missionaries and a second influenced by the rural union movement originating with Bolivia's 1952 revolution. Traditional indigenous organization from the mission period was based on a *cabildo* system introduced by Franciscan missionaries in the 19th century, consisting of several male leaders called *caciques* (chiefs), assigned by local priests. Today the role of *caciques* is largely ceremonial, though they do exert moral leadership.

Starting in the 1970s, the Guarayos people began adopting an organizational strategy to occupy and allocate land reflecting the practice of the highland rural unions, in this case referred to as agrarian zones. Small villages have a single agrarian zone, but in larger towns multiple agrarian zones come together to form what is called a *central* (again, modelled after the rural union movement). In both large and small settlements, communal assemblies headed by an elected president hold decision-making power over natural resources, allocate land to agrarian zones and mediate disputes. These village-level organizations provide the basis for the system of indigenous political power.

Lands immediately surrounding settlements are divided into agricultural zones and, beyond that, forestlands and wetlands are considered 'zones of influence', loosely defined to distinguish territories between neighbouring communities. The agricultural zones are authorized or sanctioned by the village *central* at the request of groups of local indigenous families seeking land to cultivate. Single communities may have only one such zone or 30 or more, depending on the size of the population. The agricultural zones are communal areas in which each family is granted customary ownership of a plot, typically about 50 ha containing swidden agriculture fields, fallows and forest areas. Ownership is based on use and the land can be passed to descendants but cannot usually be sold. The zones of influence are generally forest areas used by community members for subsistence (hunting, extraction) but also for the expansion of agriculture. Neither agricultural zones nor zones of influence have any formal or legal standing, though some families had received formal titles to their plots during previous rounds of agrarian reform.

Six *centrales* represent the Guarayos population in towns and small Guarayos communities scattered across the province. These organizations come together in general assemblies and elect leaders who form the core of COPNAG. In 1996 COPNAG presented a TCO demand for almost 2.2 million ha to the government covering most of the Guarayos province; the claim was reduced to 1.3 million ha after the government's spatial needs study (VAIPO, 1999). Through a rule referred to as 'immobilization', new third-party claims in the area are prohibited until titling is completed.

To demarcate the TCO, the *Instituto Nacional de Reforma Agraria* (National Institute of Agrarian Reform, INRA) evaluates competing claims through a review process and 'regularizes' property rights before issuing the land titles. Legitimate third-party claims to land within the TCO demand were made by landowners with long histories in the region and others who had previously purchased land and received title. After the settlement of competing claims, state lands would be turned over to the TCO through collective titles held by COPNAG in the name of the Guarayos people. Even though the law mandates indigenous participation, COPNAG had limited control over the decisions made by government agencies.

In 1997, for example, as the process began, COPNAG filed a legal challenge to keep the Forest Superintendence from awarding timber concessions to industries that previously had exclusive access to the region's forests. The companies held contracts valid for 20 years (set to expire in 2010) and COPNAG argued that the 40-year concessions awarded under the new forestry law constituted *new* rights in 'immobilized' areas (Vallejos, 1998). Nevertheless, later that year the Forest Superintendence rejected COPNAG's position and determined that the industrial rights were preexisting. As a result, more than 500,000 ha of production forest, most of it overlapping the TCO demand, was granted as concessions to 11 timber industries.

INRA adopted a strategy that allowed rapid progress in titling but did not address the immediate land security problems faced by most Guarayos residents. Rather than focusing at the settlement scale and addressing customary properties delineated by agricultural zones, INRA instead grouped large expanses of territory into five 'polygons', independent of the pattern of indigenous land use. INRA concentrated on remote polygons with few inhabitants first, instead of attempting to secure indigenous landholdings where most indigenous people lived. The strategy allowed the agency to cover huge territories rapidly and avoid resolving competing claims in more densely populated areas. By the end of 2003, about 1 million ha from the first two polygons had been titled; by late 2006 an additional 18,000 ha had been titled in the third polygon (only 7 per cent of the polygon's area). These titled areas are mostly remote and in some cases form irregular archipelagos of titled patches. There has been little progress in the fourth and fifth polygons that surround the highway and main town, where most of the population is concentrated.

Once the TCO demand was accepted, COPNAG was given power and administrative responsibilities over the territory – roles for which it was not designed or prepared. Mechanisms for collective decision-making, clearly defined rights and responsibilities of leaders, as well as processes for oversight by constituents, were not sufficiently developed. Representation and consultation with constituents suffered because of the distance between remote communities and a leadership based in the provincial capital – a problem compounded by weak transportation and communications infrastructure. COPNAG had been created to pressure the government to recognize land claims, but with the TCO it suddenly became responsible for representing Guarayo interests to the government, allocating resources by supporting forest management requests of indigenous residents and certifying the authenticity of preexisting land claims by non-indigenous people.

The competing claims of non-indigenous residents were a thorny issue involving economically and politically powerful individuals, not all of whom had legitimate claims. Though the 'immobilization' of the territory was supposed to freeze land transactions while the agency sorted out claims, long delays in the review process in the most contested areas (more than ten years, to date) and INRA's emphasis on titling uncontested lands allowed illicit land transactions to take place in the accessible lands that were highly prized by both indigenous people and outsiders. For example, COPNAG leaders were implicated in providing forged certification documents for landowners (López, 2004; Moreno, 2006) and charges surfaced that in 2001 there had been 44 fraudulent transactions involving private landowners, COPNAG leaders and INRA technicians (López, 2004).

The influx of economically powerful actors laying claim to large areas of land for ranching and other agro-industries had risen sharply and continued to be a source of conflict. Many had arrived to establish cattle ranches, but increasingly soybean production has become more important, requiring the transformation of extensive areas of forest. It is difficult to know the number of such properties. However, examining preliminary data from INRA, Cronkleton and Pacheco (2008b) estimate that these actors control some 20 per cent of the province.

The atmosphere of illegal transactions has also begun to undercut customary land allocation systems. For example, some families that had received individual title or other documents authorizing their occupation sold these rights to outsiders and moved further into the forest to establish new plots. In other cases, indigenous members of agricultural zones that were claimed by ranchers or non-indigenous farmers accepted payment to drop their claim to the land. The indigenous families apparently expected that large areas were going to be titled to the Guarayos people.

COPNAG has not been an effective territorial governance institution, but it is only fair to note that it faced difficult circumstances that would challenge the capacity of much more consolidated organizations. One of the problems is scale. Working at the territorial level limited the effectiveness of the TCO as a property rights institution because mechanisms for resource allocation customarily worked at the village level. Also, as an entity, the TCO is vague and incomplete, not completely contiguous, and home to a diverse ethnic mix with a significant non-indigenous population. More importantly, as has been noted with other TCOs whose property boundaries do not conform to political–administrative divisions (Stocks, 2005), the Guarayos TCO overlaps several municipal governments. These recognized political units have legal attributes, responsibilities and powers and are part of the national civic-administrative structure, independent of indigenous governance institutions.

In several ways the titling process itself inhibited the development of strong indigenous institutions and undermined the conception of the TCO as a cohesive entity. INRA's delays, and the decision to recognize timber concession rights over indigenous rights, undermined confidence that government institutions would defend indigenous interests. An impoverished population is more susceptible to influence and bribes that respond to their individual interests over the interest of the collective. Unprepared for its assigned tasks and under extreme pressure from external actors in a highly charged environment, the indigenous organization was unable to control the process, its members or its leaders.

The accusations of fraud and the influence of competing interests have continued to generate turmoil in COPNAG and the Guarayos political movement. In 2007 the organization split in two, the former leaders were expelled and a woman was elected president. The expelled leaders formed a parallel group they call the 'authentic' COPNAG, which has been recognized as the legitimate representative of the Guarayos TCO by Santa Cruz's departmental government and the *Comité Cívico* of Santa Cruz – which also represents the interests of the industrial timber sector. The original organization is divided much along the contours of the national political conflict between the central government (in favour of the indigenous president, Evo Morales) and regional departmental governments (against Morales and demanding regional autonomy). Such internal conflict further complicates indigenous efforts to consolidate the territory.

Philippines: Ikalahan ancestral domain[6]

The first indigenous community in the Philippines to receive recognition of its forest rights were the Ikalahan (also known as the Kalanguya) people, who won formal rights to use, manage and exclude third parties from the Kalahan Forest Reserve in 1974 through a memorandum of agreement with the state forest department. Formally, the reserve was public land under the jurisdiction of the Department of Environment and Natural Resources, which assigned management and oversight to the Bureau of Forest Development. Prior to the agreement, the state held all formal rights to the land and forest, but the Ikalahan people used and managed the area according to their customary practices. This included the allocation of forest areas to individual families, who could then harvest forest products, use the land for swidden agriculture and transfer their plots through inheritance or in return for payment in cash or in kind (for 'improvements') to other members of the tribe. There were no restrictions on hunting or gathering (Dahal and Adhikari, 2008), and large areas were deforested (e.g. 50 per cent and 60 per cent in two of the seven communities) (Dizon et al, 2008).

The struggle of the Ikalahan people for the formal recognition of their rights began in the late 1960s in response to outside encroachment from land grabbers. In 1968, a few prominent politicians obtained title to about 200 ha of tribal lands and in 1970 the government was planning to occupy more than 6000ha and build a vacation resort called Marcos City. In 1972 the Ikalahan won a court ruling voiding the claims of these external actors but obtained no legal document securing their own rights.

Like the Guarayos people in Bolivia, the Ikalahan decided to form an organization to fight for formal recognition of their land claim. With the assistance of an American missionary, Pastor Delbert Rice, they formed the Kalahan Educational Foundation (KEF; it was originally set up to establish a high school, hence the name). In 1974, after two years of negotiations, KEF signed a memorandum of agreement that designated the forest as a community forest for 25 years (Dahal and Adhikari, 2008). It largely recognized customary rights and practices, except for the right to transfer land among members of the tribe, and granted exclusion rights to KEF. It also established management guidelines that KEF was required to follow.

By the time the agreement expired 25 years later, in 1999, the Philippines government had passed the Indigenous People's Rights Act (Republic Act 8371), establishing the procedures by which tribal communities could obtain certificates of ancestral domain titles. This title formally recognized the rights of possession and ownership of indigenous cultural communities over ancestral domain areas to which they could prove historical possession (Dizon et al, 2008). Section 58 of the 1997 act states that indigenous communities shall be given the responsibility to maintain, develop, protect and conserve their ancestral lands with the full and effective assistance of government agencies.

KEF obtained a certificate of ancestral domain claim valid for five years as part of a community-based forest management agreement, pending full implementation of the Indigenous People's Rights Act. In 2006, it obtained its permanent certificate of ancestral domain title. The title recognizes Ikalahan rights to 14,730 ha. It confirms all of the tenure rights granted by the original agreement as well as the right to conduct internal land transfers and, perhaps most importantly, it is permanent. KEF is the formal representative of the tribe and the designated institution with decision-making power over land and forest management.

KEF has about 500 member households in seven communities (*barangays*, which are the smallest units of political administration). More than 90 per cent of the people living in the reserve are Ikalahan, and all Ikalahans are automatically KEF members. The adults in each *barangay* constitute the Barangay Assembly and are all voting members. Each *barangay* has elected local government officials (the *barangay* council), tribal elders (almost always men) and informal tribal leaders. According to Rice (2001), elders hold office by ascription and are recognized as effective at providing leadership and resolving disputes, but they do not represent the community or make decisions for the community. The most important institution is the Tongtongan. The Tongtongan functions like a tribal court, presided over by local elders, whereby the community comes together to discuss a conflict or problem; the elders make the final judgment, which is aimed at reconciliation (Rice, 1994). In effect, the Tongtongan, as an informal or customary institution, is even more important for decision-making than KEF.

KEF was formed by a group of elders and its first board of trustees was made up of one representative from each of the participating *barangays* plus an additional representative from the most populous community and one youth representative. A representative of all the *barangay* local government offices was allowed to attend the meetings but originally without voting rights. Today, of 15 voting members, three are women. The *barangays* each choose their representatives for two-year terms in general assembly meetings, which are held twice a year. Elders and older community leaders continue to dominate the board and younger, better-educated Ikalahan occupy the technical positions of the foundation. Rice (2001) believes that the elders have more effective social skills that 'keep the community members working together'; the elders also ensure that 'social problems are fully recognized in the development programs'.

Before KEF and the recognition of tenure rights, the situation appears to have been characterized by open access. The elders and local government could influence the use of land based on existing customary rules and government regulations but had neither full control over its use nor the power to exclude others, including speculators, from entering and claiming portions. With the memorandum of agreement and now with title, KEF is charged with establishing and enforcing the rules and regulations for the reserve. Today, these include regulations regarding swidden farming, tree cutting, chainsaw registration, fishing, quarrying, hunting and land claims. KEF also addresses forest fires, illegal entry and the use of sanctuaries. Bans have been established for certain tree and non-timber forest product species. The rules also establish penalties for violations. KEF approves the allocation of all household parcels by issuing certificates of stewardship contracts signed by the farmer and the board of trustees. The board must also approve land transfers among tribal members. KEF's agroforestry office provides permits for all tree cutting, which also has to be approved by the *barangay* captain, or government officer. Any clearing of new land also requires a permit from the agroforestry office.

The relationship between KEF, local governments and community members is largely harmonious. Dahal and Adhikari (2008) report that the relationship with *barangay* governments is based on trust and mutual cooperation, including shared revenue from timber permits. Community members also largely respect the rules, following the principle that all stakeholders are responsible for following and enforcing them. The rules and regulations were presented and discussed in each *barangay* before final approval by the board of trustees. There is also an incentive for catching violators: the person denouncing the violation has a right to half the funds from the fine. The regular general assembly meetings are open to all, and when important issues need to be discussed, attendance and participation are high (Dizon et al, 2008). A focus group analysis of power dynamics before and after the tenure change concluded that the farmers themselves, who had previously had less influence than other actors, were now highly influential, together with KEF, regarding forest and land management. In other words, the current arrangement is seen as giving community members greater voice.

The Tongtongan continues to be an important informal institution for problem solving and collective decision-making and works hand in hand with the KEF governance system. Honesty, equity and fairness are explicitly promoted. In one case, the chair of the board was implicated in illegal harvesting and transport of timber from the forest and he was penalized (Dahal and Adhikari, 2008). A third-party financial audit is conducted every year. Pastor Rice, who played an important role in building the 'bonding social capital' and encouraging fair internal management, serves as executive director of KEF and helps mediate relationships between the community and external actors, such as the government, donor agencies and NGOs. The situation of KEF, which has been devolved substantial decision-making powers, is not typical of other community forests in the Philippines, however.

Ghana: Traditional lands and trees[7]

Though the state legally manages Ghana's natural resources 'for the benefit of the population', especially the landowning communities, important tenure reforms have occurred since the return to constitutional rule in 1992 and adoption of a forest and wildlife policy in 1994. The constitution and this policy, as well as related laws and acts, directly promote the sharing of benefits from forest products with forest-fringe communities. These include the distribution of stumpage fees, the negotiation of social responsibility agreements (SRAs) and benefits to farmers. The first two will be considered here.

The formula for the distribution of stumpage fees, established by the constitution, mandates 25 per cent to the stool (a family or clan represented by a chief or head of family), 20 per cent to the traditional authority (presumably the paramount chief) and 55 per cent to the local government. SRAs are negotiated between logging companies and communities to provide 'social facilities and amenities' in contracted logging areas. Though these changes provide new opportunities for communities to benefit from forest resources, they are fraught with problems, including the issue of representation, especially with regard to revenue distribution.

The political structures in the local arena in Ghana involve both modern and traditional authorities. Ghana has a decentralized local political administration. The most powerful political institution at the district level is the district assembly, which has deliberative, legislative and executive powers and is made up of both elected and appointed members. Typically, each community (village or town) may be represented by one or more elected persons, called assemblymen (sometimes including women). At the community level, the main political entity is the unit committee, which includes the assemblymen and other elected and government-appointed members, all from the community.

Land and resources, however, cannot be separated from the traditional system of landownership in Ghana and the institution of chieftaincy; both are preserved in the 1992 Constitution (Art. 267). Although the state has vested control and management rights for all natural resources in the president, the ownership of these resources remains in the hands of traditional authorities. To understand the linkage between traditional institutions and rights to forest benefits under the various reforms, it is important to understand the complexity of the traditional chieftaincy system and territorial jurisdictions.

The chieftaincy position is hereditary, based on membership of a royal family or clan belonging to a community that has collective ownership of a specific portion of land, called stool land. A traditional area is an area within which a paramount chief exercises jurisdiction. That is, traditional areas are not linked with state administrative boundaries but rather associated with a paramount chief. Each paramount chief presides over two ranks of subchiefs and the elders in a traditional area's council. The paramount chiefs in specific administrative regions form the Regional House of Chiefs and five elected members from each regional house of chiefs in turn form the National House of Chiefs (Art. 271). Customarily, all the land in a traditional area is 'symbolically' under the paramount stool, but ownership is complex and subject to multiple claims, especially by the lower-level chiefs.

Chiefs are important and powerful leaders in Ghana because they have a certain legitimate claim of custodianship over community properties and rule over specific territories and domains. In short, the jurisdiction of chiefs, recognized in customary and often in statutory law, has permitted them to assume positions as community representatives since colonial times. Today, in most SRA negotiations with logging companies, chiefs represent communities and in almost all cases are the signatories. Marfo (2001) studied SRA negotiations in five communities and found that traditional leaders exerted substantial control over decision-making, and in some cases their opinions silenced the views of elected community leaders. Another study found that in five out of nine cases no entity representing community interests other than the chief was involved in the contract (Ayine, 2008). Further, in some cases, provision for marginal side-payments to chiefs and other community leaders were included in the agreement; in one case, US$600 was to be paid monthly to a paramount chief. When other political leaders are present, they are often unable to challenge the chief's position regarding the content of 'community interest' (Marfo, 2004). An attempt to follow the SRAs of all 173 licensed timber operators in Ghana concluded that even though the legal framework provides an enabling environment for negotiation, the practice of negotiating and implementing these agreements to benefit communities leaves much to be desired (Ayine, 2008).

Forest revenue is supposed to be distributed to the stool and to the traditional authority. The law is ambiguous in its details, however, stating that funds should be directed 'to the stool through the traditional authority for the maintenance of the stool in keeping with its status' (Art. 267). There is no explicit requirement that the stool's 25 per cent be reinvested in the community. Nor is it clear whether local people should benefit from allocations to traditional leaders in their private capacity.

Opoku (2006) observed that 'chiefs tend to appropriate royalties for their personal or household use and have often claimed that this is the meaning of "maintenance of the stool in keeping with its status"'. They argue that only the royalties allocated to the local government belong to communities. Nevertheless, the constitution states that 'ownership and possession of land carry a social obligation to serve the larger community and in particular,...the managers of...stool...lands are fiduciaries charged with the obligation to discharge their functions for the benefit respectively of the people of Ghana, of the stool... concerned and are accountable as fiduciaries in this regard' (Art. 37, s. 8). Also, as Opoku (2006) argues, since land is communal property, it follows that royalties belong to the community as a whole and not to chiefs. The 'status' of the stool can therefore refer only to the well-being of the community that it symbolizes; chiefs in customary law are custodians of the community interest.

The allocation of forest revenue is also contested based on multiple claims within the hierarchy of traditional authorities. Some subordinate stools complain that payment of royalties through the traditional authority enables paramount stools to appropriate part or all of these funds (Opoku, 2006). Marfo (2006) documented one such conflict between the chief of the Juaso stool lands and the Dwaben paramount chief. The former argued that he and his elders constituted the traditional authority over the Juaso stool land and thus should receive the Juaso stool revenue directly. For his part, the paramount chief argued that he owned all lands in the traditional area over which he exercised jurisdiction and should thus receive stool revenues through the traditional council's account.

Though cultural practices make it difficult to challenge paramount chiefs openly, similar complaints have been heard in private communications. Also in private communication, many chiefs have reported either that they have not received their revenue from the traditional council for some time or that the amount received may not reflect the actual amount paid. Thus, even if stool land chiefs use forest revenue for the collective interest of the communities they are supposed to represent, the traditional system may severely limit the downward flow of benefits.

In addition, by allocating 20 per cent of royalties directly to traditional authorities, the constitution further blurs the customary law distinction between ownership and political leadership. Customarily, lands belong to the stool and therefore every chief has jurisdiction over his own land. But the paramount chief, as head of the traditional council of chiefs, holds a political leadership position that does not confer ownership rights over land. Opoku (2006) concludes that the benefit-sharing arrangement condones state sponsorship of elite chieftaincy institutions in a way that gives them a stake in the system whereby timber companies exploit community resources.

The issue of community representation determines the extent to which the reforms promoting benefit sharing will actually affect the lives of ordinary people. There is increasing interest in Ghana in rethinking the community arena and building a governance culture that allows for the redefinition of community representation. Indeed, researchers have observed that community members may prefer other structures, aside from their chiefs, such as a community development committee or other elected committee, to represent them (Marfo, 2004). In other words, tenure reforms that grant rights to communities to share in benefits associated with timber present an opportunity but are effective only if representation and decision-making are accountable and can guarantee that benefits actually reach the intended recipients. Though the new policies work towards this goal, the lessons from SRAs and the distribution of forest revenue do not inspire much hope.

Role of authority in tenure reforms

The four cases demonstrate that the recognition of rights to land and forest resources is rarely straightforward, and the choice of institution representing the collective affects the rights and access to resources in practice. Even though the method of recognition (Fitzpatrick, 2005) and the role of traditional or customary authorities vary somewhat across the cases, the choice of institution is the central site of struggle and contestation.

The politics of authority plays out in different ways. Competing interest groups or institutions can manipulate the process of implementing reforms, leading to a loss of rights for intended beneficiaries. In the Nicaraguan case, this involves manipulating the definition of the territories and the choice of territorial authorities. In Bolivia, the territorial institution found its authority challenged when decision-making over what constituted a legitimate prior claim over 'original' lands was decided by the state in favour of agrarian and timber elites, culminating in the intentional corruption of indigenous leaders as part of rampant land grabbing. Ghana's traditional authorities tend to usurp benefits for their personal use. Only in the Philippines case did authority politics create legitimate power – an effective authority that was legitimate to both the state and the community. All of the cases demonstrate the central role of representation and downwards accountability in the choice of authority.

This section begins with a discussion of the domain of powers and choice of institution in each case, based on Fay (2008) and Fitzpatrick (2005), and the relation of traditional authority to the institutions recognized. It closes with a discussion of authority relations as a site of struggle.

Domain of powers and choice of authority

In each case study, the institution selected to represent the collective is granted an important domain of powers that shapes resource rights on the ground. In Guarayos, an indigenous people's organization was granted the power to support logging petitions and to certify the validity of land claims. In Nicaragua, leaders at the territory level approve logging permits and have access to tax income designated for the territory. In the Philippines, an institution headed by elected representatives grants land and forest access permits and establishes management norms defining resource access. In Ghana, traditional chiefs receive forestry funds and negotiate agreements for social amenities, both of which should benefit the community.

Ghana is the only case that uses an agency approach to representing the collective: the chief is the designated representative of the community, or stool. Fitzpatrick (2005) writes that the main advantage is simplicity but warns that representatives may not act in the interest of the group. This is the fundamental issue regarding chiefs in Ghana, who appear to be battling among themselves for access to timber funds and excluding other community representatives in the negotiation of agreements on behalf of the community. Currently, traditional authorities appear to present substantial obstacles for communities to gain benefits from the opportunities presented by new forest rights.

The Bolivia and Philippines cases both use the group incorporation approach but with different results.[8] The advantage of incorporation is that a clear, corporate structure with written bylaws and regular elections helps 'prevent internal abuses of power' (Fitzpatrick, 2005). In both cases, the organizations began more as social movements aimed at fighting for formal recognition of their communities' land and then became administrators of the territory. The Bolivian institution was less successful in making this transition, for several reasons: the Guarayos territory faces multiple and significant external pressures, the land area is almost 2 million ha and the population the organization represents is highly dispersed. In contrast, the Philippine institution manages less than 15,000ha over seven communities and a contiguous land area without competing claims. A foreign pastor with significant moral authority facilitated effective organization and communication over many years. It is important to note, however, as Fitzpatrick points out, that this structure does allow for legal recourse in the case of abuse of power, and such recourse has been taken in the Guarayos case.

The Nicaraguan case falls somewhere between these two approaches. At the community level, the law recognizes the existing communal authorities, who are periodically elected. At the territory level, the authority is supposed to be elected by the communal authorities based on an internally identified method and structure. As in the agency method, these representatives then act on behalf of the community without the formation of a new corporate structure. Unlike most agents, however, they are not appointed but elected, at least in theory. Both the communal and the territorial entities do become formal, legal institutions, hence presumably with legal recourse mechanisms – similar to the group incorporation approach, though they do not, at present, have written bylaws.

In all the cases, the institutions chosen to represent communities are based, to some degree, on traditional authorities, recognizing or taking into account existing customary arrangements and/or promoting the creation of new institutional structures based on the demands of those receiving new tenure rights. Only in the case of Ghana are these traditional leaders non-elected and hereditary. In the Philippines, tribal elders are very important in Ikalahan communities, but they hold formal office only if they are elected as community representatives, either to the *barangay* (local government) or to the institution, which in the past has included many elders. In Nicaragua's Miskitu communities, leaders are elected in annual community assemblies; a group of elders may be recognized as having certain moral authority or given a role in oversight of younger elected leaders. In Bolivia, the leaders of the institution were elected by the Guarayos organizations and communities that came together to fight for land. Though these traditional institutions may, at times, be autocratic and self-interested, abuse of power is clearly not limited to traditional authorities, and such authorities are an integral part of the most successful authority studied, the Kalahan Educational Foundation, in the Philippines.

As might be expected, the crucial variable is not whether an authority is traditional, but whether it is accountable. The Kalahan Educational Foundation was built by tribal elders with substantial support and influence from an outsider who pushed leaders to work for the benefit of all the Ikalahan people and helped establish effective accountability mechanisms. In this case the traditional Tongtongan meetings – where the entire community comes together before the tribal elders to discuss and resolve problems – are seen as playing an important role as well. In Nicaragua, conflict has arisen precisely because some communities have come together to build multicommunity territories and governance organizations on their own terms, with significant efforts, informed by past experiences, at improving accountability and representation. But then they are faced with the imposition of other territories and authorities they did not choose and who do not represent them. In Ghana, the problem resides precisely in the fact that chiefs and clan leaders are not elected, are not necessarily accountable to communities and cannot be removed from their posts.

Authority relations as site of struggle

The site of struggle through which authority relations are contested, negotiated or manipulated, and from which legitimate power may emerge, is defined by three sets of relations: between the state and the community, between the state and the institution representing the community, and between this institution and the community it is supposed to represent. Exploring these relationships helps tease out the nature of conflict (or resolution) in each of the cases. The central site of contestation in the relation between the state and the community is the *choice* of institution selected to represent the community and the *nature* of that institution, which then sets the stage for the other two sets of relationships. First, the institution chosen by the state may not be the same institution that the community has chosen. Second, this institution may not implement its domain of powers through accountable relations with the community.

In Nicaragua, communities have come together as territories and elected their own territorial authorities – only to have these ignored by government officials in the autonomous regions. Indigenous party leaders have sought to impose their own configuration of territories and virtually assign their chosen 'leaders' as territorial coordinators. In a sense, these political officials are seeking to impose an agency approach – their agent – on the recognition of local authority. Hence, the state and community are each putting forward its own authority and seeking to win legitimacy with the other. For its part, the state seeks to legitimate its chosen institution through controlled elections, political pressure, party politics and patron–client relations; communities have used multiple mechanisms, such as seeking donor funds to demarcate their own territories, electing their own territorial authorities and using their personal and political connections to get them registered.

The state is the more powerful player, however. Its imposed institutions control access to forest resources for commercial purposes and receive funds destined for the territory. Though to date the smaller-scale territorial authorities elected by the communities are still (at least in some cases) being asked permission for resource access, this structure is far from building the self-determination that many believe should be at the root of regional autonomy.

Given the proposals on the table, it is unclear how the domain of powers of these imposed territorial coordinators, which has already shifted powers away from those proposed by communities themselves, may expand in the future.

Though the history is different, the current situation in Guarayos, Bolivia, is similar to that of indigenous communities in Nicaragua, with the state – in this case the departmental government of Santa Cruz – and indigenous communities each having recognized a different institution to represent the territory. When indigenous communities initially elected an institution to struggle for the recognition of their land rights, it was a legitimate and effective social movement organization. At the same time, state authorities failed to defend indigenous territorial rights over the renewal of logging concessions, moved slowly on titling and avoided the most vulnerable areas. This led – at least in some places – to an open-access dynamic and the collusion of national-level, and some local, leaders in the sale of land rights. Severe pressure for land (with a perception that indigenous titling would not hold), a new role for which it was not prepared and ineffective mechanisms for downwards accountability undermined the legitimacy of that group of leaders.

In response, Guarayos indigenous communities, using the powers granted to them through the bylaws of their incorporated organization, ousted the discredited leaders and elected a new set of representatives. They have sought and won legitimacy through grassroots support and the investigation of the former leaders for corruption. The contest for power has continued, however: the ousted leaders formed a parallel institution that was immediately recognized by departmental officials – aligned with industrial elites – as legitimate.

The politics of authority in Ghana is somewhat different. Though the agency approach and the choice of institution representing communities are contested, there is also considerable debate regarding the nature and domain of the institution that has been recognized – in this case the traditional authority – and, therefore, regarding the relationship between the community and this institution. Traditional institutions in Ghana have been recognized as the main recipient of stumpage fees on behalf of communities and have assumed the role of community representative with logging companies in negotiating agreements for social amenities. Their presence in negotiations appears, in general, to silence other community representatives, including those who were elected by the community. Evidence so far regarding timber fees suggests that few of these benefits are likely to reach communities.

Some chiefs at least do not believe that they have any obligation to the communities they represent, and the vague wording of the legal statutes gives cover. At the same time, the constitution clearly establishes that chiefs are fiduciaries with a social obligation to the communities for whom they hold land. The state has explicitly granted a portion of funds to customary leaders and considered them legitimate community representatives but failed to hold them responsible for investing timber revenues in the community. For their part, different levels of leaders within the traditional structure also claim the right to be recognized as legitimate recipients of fees – but not, apparently, to be more accountable community representatives.

Finally, in the Philippines, the state and community have both agreed to recognize the same institution and hence its legitimate power as the representative of the Ikalahan people. The Kalahan Educational Foundation has been successful in part because it was built on a foundation of honesty and equity, with effective mechanisms for participation and accountability. Moreover, the ethnic community is homogeneous and an external participant has been an informal but influential moral authority, negotiating both upwards and downwards accountability and acting as a cultural broker. Dahal and Adhikari (2008) credit the president of the institution, Pastor Rice, with catalysing high 'bonding social capital' (for internal cooperation) and promoting fair internal management practices; they believe that his relationships beyond the community also helped support two other types of social capital – for bridging social differences and divisions and linking to people in power.

The Kalahan Educational Foundation maintains respectful coordination with the local elected governments, which are seen mainly as providing government services and hence addressing a different and non-competing domain of power and decision-making. Traditional elders are elected to positions alongside younger and sometimes better-educated people. Both the foundation and the elected local governments turn to the traditional Tongtongan for conflict resolution when needed. The integration of a new entity with traditional institutions and actors allows for an effective balance between formal statutory and customary practices and a high level of local legitimacy. Farmers believe that the recognition of their tenure rights has given them greater voice in resource decision-making both as resource users and as members of the foundation.

Conclusion

This chapter has demonstrated that the institution selected to represent the collective in forest tenure reforms plays a significant role in the distribution of forest rights and benefits. The four cases show that simply choosing the correct, downwardly accountable institution may not often be an option; in fact, in none of the cases did such an entity exist at the scale required. Rather, authority relations constitute sites of negotiation and conflict, and the politics of authority may play out in different ways.

Constructing legitimate power, whether based on customary or new institutions, does not occur in isolation but is based on the interactions among the state (in its multiple manifestations), the community and the institution recognized by each of these as the community representative. In two cases, the state and community recognized different institutions as legitimate. In three, the entities selected by the state were not accountable to the communities they were supposed to represent. The only case in which a single authority emerged that was legitimate both to the state and to the community involved an embedded external broker.

That finding should suggest not that external actors are a necessary condition for the construction of authority, but rather that effective representation requires transparent rules of the game, including broad agreement on how representatives are chosen, the creation of accountability mechanisms and the specific domain of powers of each authority. Communities and their organizations will need allies, both in and outside state institutions, when the state is, or appears to be, complicit in backtracking on the rights that have been won through tenure reforms. Greater recognition of the need to address these concerns and explicit emphasis on their effective resolution should lead to more positive outcomes in the implementation of new community rights.

Notes

1. For the purpose of clarification, we have tried to avoid using 'authority' to refer to the institution in power (or domain of power), but because of the widespread use of 'traditional authority' and 'customary authority', it is difficult to avoid without creating further confusion. If authority denotes legitimacy, then the term itself assumes legitimacy where this may not exist.
2. See also Larson and Mendoza-Lewis (2009).
3. Abbreviated from the Law for the Communal Property Regime of the Indigenous Peoples and Ethnic Communities of the Autonomous Regions of the Atlantic Coast of Nicaragua and the Bocay, Coco, Indio and Maíz Rivers.
4. The autonomous regions are referred to by residents as the Caribbean Coast of Nicaragua, though only part of the region is actually coastal. Hence the term is capitalized to refer to the area as a whole.
5. See also Larson et al (2008), Pacheco et al (2008b) and Cronkleton and Pacheco (2008b).
6. Special thanks go to Josefina Dizon, Tamano Bugtong and Mona Pindog for helping clarify this section.
7. See Marfo (2009).
8. The Bolivian state allows organizations like COPNAG to write their own bylaws based on *usos y costumbres*, or customary practices; that is, it does not impose or require any specific internal structure.

6

Community Networks, Collective Action and Forest Management Benefits

Naya S. Paudel, Iliana Monterroso and Peter Cronkleton

This chapter focuses on understanding the forms, scope and nature of secondary-level community institutions and their role in securing tenure and enhancing forest-based benefits. The previous chapter has made clear the problems that remain in efforts to construct fair mechanisms for representation and devolution of appropriate authority. Here, we take a pragmatic view, looking at collective action at a secondary level, such as networks, and explain the role they play in mediating the reform processes and ultimately shaping their outcomes.

Community networks build on community-level collective action, scaling up these actions at higher levels. *Community network* (hereafter 'network') refers to the diverse forms of secondary-level organizations of forest-dependent communities, such as federations, alliances and associations that defend and promote community interests. Despite the large literature on collective action at the local level (Ostrom, 1990,1999; Agrawal, 2001; Bromley, 2004), scholarly work on secondary-level organisations, particularly networks, is relatively scarce. This chapter explores the evolution and dynamics of networks – as sites for collective action – of forest-dependent communities and explores their roles in enhancing forest tenure security and livelihood benefits.

Networks have often become part of the forest tenure reform process. In most of the cases studied, networks emerged from a major reform process and later became the promoter of further devolution. More than isolated and localized traditional institutions, networks can be seen as modern institutions for collective action that facilitate learning, sharing, mutual exchange and collaboration. The frequent interaction and communication among local groups, in turn, help networks advance their agendas.

Since networks have increasingly become part of tenure reform processes, more knowledge and reflection about the evolution and dynamics of networks could facilitate the tenure reform process itself. What are the political and policy contexts in which such networks emerged? What are their organizational structures, scope, priorities, strategies and activities? What roles have networks played in shaping forest policy and practice? What are the major achievements in terms of securing tenure and enhancing the benefits of forest management? What are the major challenges? What theoretical and practical lessons can be drawn from these stories? The answers to these questions may expand our knowledge of community networks in the context of forest tenure reform.

The chapter explores how local communities and concerned groups of citizens can effectively participate in constituting networks and influence management practice at different scales of forest governance. We identify the conditions under which these specific networks evolve, grow and serve the above-mentioned functions. We primarily draw on three cases: the Federation of Community Forest Users, Nepal (FECOFUN), the Association of Forest Communities of Petén, Guatemala (ACOFOP) and the Integrated Agroextractivists of Pando Farmers' Cooperative, Ltd., Bolivia (COINACAPA), here referred to as the Brazil Nut Producers' Cooperative. The first two networks consist of community-level organizations involved in forest management. They are primarily involved in policy advocacy in securing community rights and enhancing community interests. The third is a cooperative of individual producers that works to secure its members' interests in the market. In addition, we also refer to similar initiatives in other countries to substantiate the discussion.

The second part of the chapter discusses theories of collective action and community networks. The three cases are presented in the third part. The fourth part, drawing from the cases, identifies some common patterns and extracts lessons regarding the evolution and dynamics of networks and their roles in securing forest rights and enhancing livelihood benefits.

Collective action and community networks

A few scholars, responding to the increased interest in secondary-level organizations, have begun focusing on the emergence and dynamics of community networks (Cronkleton et al, 2008; Ojha et al, 2008). Theorists have focused on collective action for the management of the commons at the grassroots level (Baland and Platteau, 1996; Ostrom, 1999) but have not fully explained the dynamics of networks at a higher level. We explain networks' collective action through a social movement perspective. Although the notion of social movements often indicates resistance, reflecting grievances around perceived injustices, we take a broader definition where such movements constitute the pursuit of alternative agendas, such as establishing cooperatives and taking other affirmative actions aimed at improving livelihoods. Similarly, although *network* connotes the notion of horizontal relations, here we are talking about secondary-level organizations of community institutions that exist within a hierarchy, albeit one that is less rigid than state bureaucracies or corporate bodies.

Because we take a social movement perspective to understand and explain collective action at the level of networks, we differentiate them from traditional movements. Scholars often distinguish between traditional collective action and contemporary social movements, even though these share some common features. First, social movements do not constitute fundamental (economic) classes; instead, they are aggregates of various social groups (Offe, 1985, p831). In case of the common pool resources, people from different classes may share common concerns and form alliances to promote their collective interests. Second, whereas trade unions and other political organizations often aim to capture state power, forest dwellers, landless people and ethnic minorities simply demand more secure rights over the resources on which they rely (Hickey and Bracking, 2005). Third, a corollary to the second, new social movements have shifted away from the realm of state and political parties and operate within civil society by creating 'new spaces and solidarities' (Cohen, 1983, p106). The network-led movements are neither guided by grand ideological positions nor intended to rule the nation, though they may seek more autonomy at sub-national or local levels. As Harvey (2003, p182) states, the traditional trade union movements are understood to be resistant against 'accumulation through exploitation', whereas network-led movements are targeted against 'accumulation through dispossession'. Fourth, unlike many traditional movements, which are primarily concerned with material production and distribution, indigenous people or other forest dwellers are also concerned with symbolic capital, such as identity (Habermas, 1981, p33).

Community networks have coevolved with extension of the collective action spreading across a large territory. Collective action aimed at social change can be considered a social movement (Touraine, 1985; Neidhardt and Rucht, 1991; Jelin, 1986). In fact, as popular collective action gradually transforms the participants into social actors, those actions take the form of social movements. Since these actions largely operate in the triangular space between family, state and the market, they can be identified as civil society networks (Habib and Kotze, 2002, p3).

However, these distinctions are made largely based on experiences of social movements in developed countries, particularly in the west. This Eurocentric interpretation of social movements may not adequately explain the social movements in developing countries. Forsyth (2007), for example, based on his study of environmental movements in Thailand, suggests that social movements in developing countries carry a relatively stronger class flavour. Unlike in post-industrial societies, poor and marginalized communities in developing countries have always been struggling to secure access to a livelihoods resource base. Protecting access to forests, land and water has remained the major driver behind the Chipko and Narmada environmental movements (India) and the rubber tappers' movement (Brazil). Therefore, it can be argued that in tropical forested countries, community networks – the leaders of forest-based social movements seeking resource rights – may also share some elements with conventional movements.

What kinds of socio-political environments are conducive to collective action at higher scales? First, the emergence of social movements largely depends on the political opportunities that may facilitate or inhibit collective action. Social movements emerge out of political opportunities, which are then expanded by movements themselves, creating further opportunities for new movements (Ballard et al, 2003, p3). The political regime and cultural traditions in any society may facilitate or inhibit the legitimate forms for voicing grievances. Generally, such movements flourish in a relatively liberal political regime where basic citizen rights are respected and free media function. Similarly, the greater the spatial and functional decentralization of a given political system, the more effective will be the social movement (Ash-Garner and Zald, 1987, p310). This allows more space for the lower units of government and community initiatives and gives rise to local groups of diverse nature, form and scope. Consequently, secondary-level organizations like networks, cooperatives and alliances emerge and prosper. This equally applies in forest-based rights movements. For example, Cronkleton et al (2008) observed that areas with minimal state presence provided conditions conducive to forest-based social movements because public institutions that could defend local rights or interests were absent. They found that local communities' common understanding of the threat to their collective livelihood interests served as the primary driver of these movements.

Second, how communities mobilize their resources is an important aspect of understanding these networks. *Resources* here comprise grassroots political constituents, enthusiastic local cadres and sympathetic supporters as well as material resources. Community networks place their resources under collective control for pursuing their collective interests. The networks mobilize resources and influence other groups to contend for power (Tilly, 1978, p78). McCarthy and Zald (1977, p1215), for example, suggest that the leaders act as 'issue entrepreneurs' by constructing issues and grievances. Charismatic leaders identify and define grievances, develop a group identity, devise strategies and mobilize the members, often taking advantage of political opportunities. Similarly, external support, particularly during the initial stages, is important for the emergence of such movements. Cronkleton et al (2008) found that external support in the form of official technical assistance and community funding for institutional growth remains instrumental in expanding and sustaining a movement.

Based on those general theoretical understandings of collective action and social movements, we seek to explain the networks of community groups in managing forests. We analyse cases from Nepal, Guatemala and Bolivia to explain the emergence, functioning and outcomes of community networks.

Nepal: Federation of Community Forestry Users, Nepal

The Federation of Community Forestry Users, Nepal (FECOFUN), the largest civil society organization in Nepal, represents forest-dependent communities. It emerged along with the growth of community forestry, particularly since the 1990s. Today it represents more than 14,000 community forest user groups

(CFUGs) spread across the country, which manage about 25 per cent of the country's forests (Dahal and Chapagain, 2008; Ojha et al, 2008).

Nepal introduced community-based forest management in the late 1970s to halt ongoing deforestation and degradation, particularly in the midhills. The programme gained momentum after the Panchayat system (a non-party political system under Nepal's monarchy) was overthrown in the 1990s and a multiparty parliamentary system was established. The new Parliament endorsed the Forest Act of 1993, allowing district forest officers to hand over part of the national forests to identified user groups (HMG/MoLJ, 1993). The act recognized CFUGs as self-governing, independent, autonomous and corporate institutions that could acquire, possess, transfer or otherwise manage their own property, such as forest resources or any related funds (HMG/MoLJ, 1993, Art. 43). A CFUG is a collective entity representing every household in the neighbourhood of any specified forest patch, usually through household heads as general members. The members, through their annual general assembly, elect an executive committee to carry out everyday forest management and associated activities. These communities range from ten households (managing 0.5ha of forest) to 10,000 households (managing 8000 ha). Following the new act, thousands of CFUGs formed and began to manage community forests (see Chapters 7 and 8).

FECOFUN, as a secondary-level network, emerged as community forestry developed. Four factors in particular contributed to its establishment (see Box 6.1). First, during the early years of community forestry, CFUGs faced institutional and technical challenges, such as forming effective executive committees, preparing group constitutions and forest operational plans and carrying out recommended forest management activities. They sought to benefit from exchange and sharing with other CFUGs who were facing similar challenges. Second, although the parliament endorsed the Forest Act of 1993, the government, particularly the forest bureaucracy, undermined its spirit by developing restrictive regulations and using discretionary power (see Chapters 7 and 8). The CFUGs consolidated their resistance by strengthening FECOFUN. Third, the new multiparty political system promoted democratic values, norms and principles and provided space for diverse forms of citizen groups to flourish. Fourth, external aid supported CFUG networking to promote and institutionalize community forestry. The Ford Foundation and bilateral forestry projects of the Swiss, British and Danish governments funded more than 95 per cent of FECOFUN's costs during its early phases (FECOFUN, 1999). Today it has 73 district chapters, 560 range post chapters and 11,700 CFUGs as formally registered members.

Actions and strategies

During the early 1990s, CFUGs urgently needed help with institutional and technical aspects, but the service provided by the government and some bilateral forestry projects was far from adequate. FECOFUN emerged to support preparation of operational plans, proper recordkeeping, improved

Box 6.1. Institutional evolution of FECOFUN

July 1992: First sharing workshop among CFUG representatives of Dhankuta district
February 1993: National workshop with 40 CFUGs from 28 districts
May 1995: Formation of *ad hoc* committee of FECOFUN following national CFUG workshop
September 1995: Formal registration of FECOFUN in Kathmandu

Source: FECOFUN (2002)

forest management practices and compliance with local and global standards for sustainable management (such as those of Forest Stewardship Council).

As community forestry expanded from a few intensively supported and carefully designed groups to a large number of groups across the country, some cases of unsustainable harvesting (Luintel, 2002), misuse of funds (Gentle et al, 2007) and exclusion of marginalized people (Agarwal, 2001; Nightingale, 2002) became apparent, especially in the *terai,* the southern lowland with dense and valuable forests of sal (*Shorea robusta*). The forest authorities overreacted and introduced restrictive policies for forest management, harvesting, sale of forest products and financial management. A series of government policy decisions, guidelines and circulars increased forest officers' discretionary power to sanction, monitor and approve CFUG activities. FECOFUN saw this as undermining local autonomy and resisted these 'regressive' moves. Table 6.1 lists the major government decisions and FECOFUN's responses, which are mostly targeted towards protecting community rights over forests.

FECOFUN uses diverse strategies and tactics including sit-in protests, street rallies, blockades of government forest offices, memoranda, press conferences, mass meetings and media campaigns. Apart from launching advocacy campaigns, it has fought several legal cases on behalf of member CFUGs, defending their autonomy. For example, in the fiscal year 1999–2000, it filed 15 cases in various courts. It also engages in constructive policy dialogue by participating in various policy forums, such as task forces and working groups like the national-level Forest Sector Coordination Committee and the District Forest Coordination Committee.

Achievements

The FECOFUN-led movement has had some successes (see Table 6.1). Although unable to reverse all undesirable decisions, the movement has helped stakeholders pursue collective action in forest management. For example, FECOFUN has promoted 50 per cent women's representation and greater allocation to livelihood activities that benefit the poor – ideas institutionalized by the government's 2009 community forestry guidelines. FECOFUN is now

Table 6.1. *Government action and FECOFUN response*

Date	*Issue*	*FECOFUN activities*	*Outcome*
1998	Timber Corporation of Nepal granted monopoly over timber trade	Organized street protest, held press conference, informally lobbied stakeholders	Government changed its decision; by implication, parastatal monopoly would not control trade in timber from forests managed by communities and local governments
1999	First amendment of Forest Act of 1993 gave more power to forest officers and restricted CFUG rights, required 50% of CFUG funds be invested in forest management	Organized disobedience, raised awareness in CFUGs about impacts of amendment	Partially successful; both forest officers and CFUG members can take action against CFUG committees for mismanagement, 50% requirement reduced to 25%
2000	Circular restricting community forestry in *terai*	Held mass meeting and press conference, submitted memorandum, lobbied policy-makers	Ban on community forestry handover in *terai* was lifted
2002	Forest Regulation of 1995 amended to remove provision giving 'special priority' to community forestry over other management options	Lobbied against decision, conducted nationwide campaign, held mass meetings at local level, submitted memorandum	Unsuccessful; amendment allows government to hand over part of national forest to private companies without prioritizing community forests
2003	Financial ordinance imposing 40% tax on CFUG forest product sales	Conducted nationwide campaigns, street protest and mass meeting, lobbied decision-makers	Tax reduced to 15% and limited to sale of two products (sal and sisso timber)
2005–2006	Illegal interference and seizure of CFUG bank accounts by zonal administrators	Held rallies in several districts	Seizure of bank accounts halted
2006–2007	Some community forests used by government for army barracks and Maoist rebels' cantonments	Submitted memorandum listing alternative options	Most community forests returned to local control

Source: adapted from FECOFUN (2002)

recognized as a major actor in the forest policy process. Its nationwide network and the sheer mass of people it represents have helped it to challenge power imbalances between the forest bureaucracy and local communities and increase user groups' sense of security over their forest rights.

Challenges

FECOFUN has faced both institutional and programmatic challenges. Grooming of new leaders, especially women and marginalized people, appears inadequate. Upper-caste males still dominate the leadership. FECOFUN is often criticized for its blind support of a populist agenda without proper reflection. According to some critics, it often reacts defensively to proposals rather than developing thoughtful positions (Ojha et al, 2008). A major threat is that because the leaders and cadres are affiliated with political parties, the movement could be coopted by a party agenda at any point. Also, FECOFUN is under pressure to meet growing demands by CFUGs for diverse types of services for their effective operation.

Financial sustainability is another major challenge. General institutional funding is shrinking and project-based funding is restricted to specified activities. This limits FECOFUN's ability to defend the user groups' rights. Moreover, not all the projects are compatible with its priorities (Ojha et al, 2007; Timsina, 2003). For example, when FECOFUN implemented a project on reproductive health with support from the United Nations Population Fund, many members questioned the link with forest rights.

Guatemala: Association of Forest Communities of Petén

As part of an effort to recognize the importance of forest biodiversity, the Guatemalan government, supported by international conservation organizations, established the Maya Biosphere Reserve in 1990 (National Decree 5-90), the largest protected area in Mesoamerica, encompassing more than 2 million ha (see Chapter 3). By the mid-1990s, with the civil war[1] winding down, the Guatemalan government faced a new and unpredictable conflict in the Petén. In a region characteristically lacking formal channels of communication and minimal mechanisms for governance, conservation agencies made little effort to reach out to distant and atomized forest community settlements (Sundberg, 1998). The regional economy had been based on the extraction of timber and diverse non-timber forest products. In 1994, the government, with the strong backing of USAID, legalized a formal community concession system in the multiple-use zone of the Maya Biosphere Reserve. This would be the largest community forestry endeavour implemented by the Guatemalan government.

According to Monterroso and Barry (2008), concessions became the compromise solution to establish a system of control in which all parties collaborated. The Association of Forest Communities of Petén (ACOFOP), as the representative of the community concessions, played a major role in promoting their economic, environmental and political interests. Additionally, stakeholders believed that concessions based on timber management would ensure better short-term economic benefits for residents than extraction of non-timber forest products (NTFPs) or tourism. Both the forest authorities and the community concessions saw the timber schemes as providing sufficient incentives for local participation and eventually long-term sustainability. The model sought to promote sharing of decision-making and benefits between local communities and the government, with decentralized responsibilities and rights.

A community concession consists of a 25-year contract between an organized and legally recognized group and the Guatemalan government; the contract grants usufruct rights to the former to manage renewable natural resources in protected areas.[2] Two types of community concessions can be identified: one type of organization is embedded within a community and is located inside the reserve's multiple-use zone; the other is an organization whose members belong to one or more communities living outside the multiple-use zone.

The initial concessions offered to communities were too small to be economically viable. The communities 'pushed back' to increase the size, levels of access and extent of control but needed to integrate their demands into a common discourse, since they were dispersed throughout the forest. While some communities staked claims to *maintain* rights held informally over NTFPs and to defend their customary rights of residence in the forest, other groups living outside the multiple-use zone made similar claims to *increase* access to forest resources. With the establishment of the concession system, the community groups began to form a secondary-level organization that consolidated community bargaining power.

Origins and evolution

Efforts to create a network started in 1995, when the Consultative Council of Forest Communities of Petén was established by community leaders with the support of the Rubber Tappers Union. The intent was to establish an organization that could integrate the different community claims into a single, unified voice. This organization would work towards expanding community concessions across the multiple-use zone within the Maya Biosphere Reserve. The organizers proposed to enlarge the initially small areas. Additionally, they negotiated a change in the framework to include the allocation of rights to communities outside the multiple-use zone. Above all, this network was the key to ensuring that community groups were participating actively in decision-making processes. By 1997 the network had become ACOFOP, with 22 community groups as members.

The first concession, granted in 1994, was allocated to a community group for 7000ha; the last community concession was allocated in 2002. The largest concession is 93,000ha. All told, the 12 community concessions in the multiple-use zone encompass more than 400,000ha, 96 per cent of which has been certified by the Forest Stewardship Council (see Chapter 3). Direct beneficiaries include more than 2000 families in 16 communities and three municipalities.

Activities and strategies

Today, ACOFOP defines itself as the representative of community organizations and acts as a vigilant advocate for community claims (see Table 6.2). Initially, the network focused on defending community groups' access to concessions;

Table 6.2 *ACOFOP activities*

Event	*Actions*
Contract negotiation (1994–2002)	Channelled technical, organizational and legal assistance to help communities understand implications of contracts, facilitated negotiation processes
New legal norms for integrated management and NGO accompaniment	Challenged allocation of integrated management rights over both timber and NTFPs Challenged role of NGOs as legally designated technical assistance providers and required cosigners for valid contracts Strengthened bargaining power of community organizations to select external organizations to assist them
Strengthening of community member organizations (1994–present)	Promoted legalization of community organizations and compliance with contracts Maintained communication and dialogue with grassroots member organizations through workshops, discussions, training processes
Expansion of petroleum concessions for exploration (1998); expansion of Mirador basin project (2002–2005)	Mediated and supported community concession organizations in negotiations with project promoters Engaged with communities in discussing alternative mechanisms and legal and project proposals that could benefit management of multiple-use zone Maintained communication between local and national levels on experience of community organizations in Petén through press releases, TV, radio Defended exclusion rights of community concession organizations when challenged (petroleum, Mirador tourism project) Established strategic alliances with government officials, NGO representatives
Regional development plan for Petén	Participated in multisectoral discussions for development of Mirador Park (known as 4-Balam initiative) Developed proposals to engage in project

Source: ACOFOP-CIFOR (2007)

now it concentrates on political advocacy to ensure that community concession rights are respected and that external interests do not encroach on their rights.[3] ACOFOP also provides technical assistance and *acompañamiento* (such as political and administrative advice) to member communities, strengthening members' organizational, technical and productive skills to facilitate self-management. It has facilitated access to credit and forest markets, improving members' livelihoods. Major decisions are taken in general assemblies that meet once a year. Participants appoint the board of directors together with the technical office (composed of several community technicians) to implement projects financed with members' support or by donors. The board of directors consists of seven members, selected from the legal representatives among member organizations, plus one representative of individual members, elected for a two-year period.

Achievements

The resulting arrangement among conservation authorities, community concessionaires, the timber industry and local government is a unique experiment for Latin America (Monterroso and Barry, 2008). The Petén has become a centre for learning for other community leaders. Outside Mexico, nowhere in Latin America has such a large bundle of rights to land and forest resources been transferred in such a short period of time and at the same time received government and donor investment and support. Following Monterroso and Barry (2008, 2009) and Barry and Monterroso (2008), major outcomes emerging from the work of ACOFOP include the following.

The concession model allowed communities to secure their residence in the area and hold usufruct rights for at least 25 years, with the possibility of renewal. Their members could now begin to exploit forest resources under sustainable use criteria, with standards and indicators elaborated for different resources. Community organizations and ACOFOP had to increase their capacity and project their agendas nationally and regionally to meet the challenges.

Illegal logging and the sacking of archaeological sites diminished significantly. Forest cover has been maintained, particularly compared with neighbouring protected areas where deforestation has been increasing (Bray et al, 2008; Monterroso and Barry, 2009). Community members established local governance systems based on an expanded set of rights of access, use and decision-making over their natural resources. This included organizing constant vigilance and patrol of the boundaries of the concessions.

Community concessionaires have increased incomes notably as they reap the benefits of harvesting high-value timber; such management activities provide employment in the region, directly involving 2000 families and benefiting more than 3000 families indirectly (Monterroso, 2007). More than 50 per cent of the 17 timber species managed are exported; 70 per cent of the production was value-added sawn wood. Eight community concession groups have bought their own sawmills. Through the work of ACOFOP, the annual timber management plans were accepted by local banks, ensuring access to credit.

Through the charismatic leadership of ACOFOP, it was possible to integrate dispersed local organizations into a single representative body, allowing for external advocacy and providing a vehicle to take concessions to scale. ACOFOP has made it possible for local organizations to defend exclusion rights.

Challenges

ACOFOP needs to be able to respond to the changing dynamics in the Petén and satisfy the diverse demands of member organizations. From a technical perspective, it must help its members meet the standards and comply with the regulations for sustainable forest management to renew their concession rights. Some community concession groups have limited capacity to meet these criteria and face huge transaction costs. They need support in obtaining the annual evaluations for forest management certification and for development of management and annual operation plans.

Political advocacy remains relevant: the community concession groups are constantly struggling to defend their exclusion rights to their concession areas. The demand for land is high and constant monitoring and lobbying are required. The community concessions are largely dependent on the government's political will and face constant threats that may weaken existing rights. The task of community networks goes far beyond the physical role of defending borders. It implies sophisticated and healthy levels of representation with capacity for interpretation and communication to its members about threats to tenure and resource rights. The advocacy campaign involves engagement in constructive dialogue with outside stakeholders. Proactive engagement requires money, mobilization and time. In addition, ACOFOP needs to enhance its legitimacy and credibility through democratic representation and increased accountability – major institutional challenges (Monterroso and Barry, 2008).

Bolivia: Brazil Nut Producers' Cooperative

In Bolivia's northern Amazon, grassroots organizations of rural forest peoples have been instrumental in gaining greater control over forestland and capturing a greater share of benefits from the sale of non-timber forest products. In particular, the Brazil Nut Producers' Cooperative (COINACAPA) has helped members negotiate a better position in the international market for Brazil nuts. COINACAPA has increased the bargaining power and incomes of these forest producers through fair trade and organic markets. This case describes the emergence of COINACAPA and its strategies, activities and livelihood outcomes. Whereas the cases from Nepal and Guatemala have strong elements of a resistance movement for protecting rights over resources, this is a case of collective action to increase benefits from forest management following on a grassroots campaign to gain control over forest resources. It shows how the political strengths of community networks are being used to enhance forest-based livelihoods.

Pando's Brazil nut industry

The Pando department's 63,827km^2 territory is one of Bolivia's most remote frontiers. Pando's population is small: only 52,525 people in the last census, with a density of 0.82 inhabitants per km^2 (INE, 2002). Sixty per cent of Pando's population is rural, and seasonal migration related to forest extraction produces dramatic shifts in the rural population.

Forest is the dominant land cover in the department and non-timber forest products rather than timber have provided the basis for rural livelihood strategies for generations. Brazil nuts (*Bertholletia excelsa*) have been the principal NTFPs since the mid-20th century, when they began to replace rubber. In fact, the Brazil nut has become the economic foundation of the region (Stoian and Henkemans, 2000).

During the first five years of this century, Bolivia accounted for more than 50 per cent of the world's Brazil nut exports and 70 per cent of the world's processed shelled nuts (FAOSTAT, 2007). Rural labourers, however, have enjoyed little benefit from this booming business. Historically, the estate owners, known as *barraqueros,* had control over forest resources, marketing networks and credit. In recent decades they have been joined by Brazil nut-processing plants, large, capital-intensive enterprises that have allowed the region's Brazil nut sector to thrive. The processing industry makes a significant contribution to Bolivia's forest export earnings (Cronkleton and Pacheco, 2008a), but collection is still labour intensive, and nut gatherers remain some of the region's poorest residents.

The land reform process has been contentious for most of the past decade (Ruiz, 2005; de Jong et al, 2006). However, since 2000 rural communities have gained recognition of their forest rights and some community producers have been able to organize innovative cooperative models. Consequently, the *barraqueros* have lost considerable influence. Their property rights over traditional forest estates have not been recognized and in many places they have lost monopoly control over markets, credit and rural labour.

Pando's agro-extractive communities have traditional property rights that evolved as ethnically mixed groups of peasant workers took control of forests and began working independently to extract and commercialize NTFPs. The basic production unit is the household, so initially rural families were dispersed throughout the forest to facilitate the daily extraction of wild rubber. Later, after the collapse of rubber prices, households began moving to more nucleated settlements, occupying forest holdings only during the Brazil nut harvest, January to March.

Recognition of agro-extractive communities

Bolivia's 1996 land law, known as the INRA Law, did not bring immediate change to the region; instead, a tense standoff between *barraqueros* and community producers and their representative organizations ensued (Larson et al, 2008). Initially, *barraqueros* used back channels in the Bolivian capital to promote decrees that would have created 3 million to 3.5 million ha of NTFP concessions benefiting about 200 *barraqueros* (Aramayo Caballero, 2004; Ruiz, 2005). News of these decrees catalysed opposition and a coalition of regional peasant and indigenous organizations took shape; it included the Peasant Federations of Pando, Madre de Dios and Vaca Diez and the Union of Indigenous People of the Bolivian Amazon (CIRABO), together with regional NGOs. This coalition formulated a grassroots response to put pressure on the national government, which was increasingly interested in populist measures to appease rural tensions. The government eventually decided that the minimum area provided to farming and indigenous communities in Brazil nut territories would be 500 ha per family (Ruiz, 2005). The measure corresponds roughly to the area traditionally used by extractivist families to harvest NTFPs and effectively recognized their *de facto* hold over extensive forest properties. However, rather than attempting to title individual plots, the policy was interpreted such that communities would receive communal properties more or less equivalent to 500 ha per family. This has resulted in the titling of

1.8 million ha of forest in Pando to only 163 communities (Cronkleton and Pacheco, 2008a).

The focus on communities for titling purposes made use of traditional institutional frameworks. These loosely organized groups of families share common claim to forest, but individual households form the basic production unit. Access is organized at the household level to manage forest holdings and to link with markets and informal credit sources. There is episodic collaboration among residents, usually along kinship lines. Communal authority, usually derived from informal collective consensus, allocates individual rights, based on customary practices that determine who has legitimate rights to use resources and where. Working alone, households had little leverage to negotiate with buyers, having to accept terms of trade that provided low prices, which often left the families in debt.

Formation of COINACAPA

COINACAPA is a small producer group that has gained access to the fair trade market. Formed in 1998 as a cooperative, it first exported a half-container of nuts (8 tons) in 2000. Its strategy is to subcontract one of the region's processing plants to shell members' Brazil nuts, which the group then exports directly to fair trade brokers in Europe. The intent is to support small producers rather than the processing plants that usually act as intermediaries exporting the nuts. By selling directly to overseas buyers, COINACAPA members receive almost twice the local market price for Brazil nuts they deliver to the cooperative. Since achieving fair trade status in 2001 it has used its premium to provide health care and other services for its members. As a result, its membership has grown from 41 families in 2001 to 465 families in 40 agro-extractive communities in 2007. By 2007, it exported seven containers totalling 112 tons of shelled nuts per year.

COINACAPA leaders say that using the market mechanisms of organic certification and fair trade arrangements has had more influence on management and production practices than any norms or forest policies issued by the government. For example, to qualify for these programmes, COINACAPA members must maintain quality-control standards for sanitation, humidity and safe post-harvest storage and transport to ensure that the nuts are free of chemicals, fuels and other contaminants. The members are organized into groups of four or five producers at the community level to ensure compliance. If nuts spot-checked at delivery fail inspection, the lot of the entire group is rejected, which creates a strong incentive for self-regulation. To demonstrate that they are small producers, members must map and document the location and size of their Brazil nut groves (measured in number of trees), which also allows better planning. By increasing benefits from forest products, COINACAPA is creating incentives for members to maintain natural forests.

Community networks as emerging actors in forest management

Some common patterns have become clear from the cases discussed above. Diverse forms of community networks have been emerging in several countries. We identify three dimensions of the community networks: conditions for emergence and evolution, strategies of resource mobilization and outcomes in securing resource tenure and livelihood benefits.

Emergence of community networks

Collective action at higher levels tends to emerge because of a perceived crisis in access to valuable resources. Both the forest communities in the Petén and the agro-extractive communities in Bolivia organized to defend their rights when they saw their interests at risk. Communities in the Petén had harvested NTFPs before the establishment of the Maya Biosphere Reserve and some people were employed by the logging companies. The establishment of the reserve restricted logging activity, directly affecting these communities. In Bolivia, indigenous communities and other small producers found themselves at risk when the *barraqueros* sought large NTFP concessions. In Nepal, community forest users wanted to share and learn about group organizing and active forest management; when their rights and autonomy were undermined, they consolidated their resistance against government decisions through the federation movement. All these networks grew out of the people's struggles for the right to maintain their livelihoods. Their current activities, however, cover other aspects, including trade, enterprise management, equity and health.

Besides the three major cases presented here, two examples from Cameroon (see Box 6.2) and the Philippines (see Box 6.3) also show that forming community networks has become a way to protect community interests in forest management.

> **Box 6.2 Cameroon:**
> **Forest management associations**
>
> To pull together local community forests management efforts, intervillage associations have been set up. In the Oku area, the Association of Forest Management Institutions was established in early 2000. In the Kirby area, a union of community forests was formed in 2007, with logistical support from a coalition of organizations, including Planet-Survey, World Wildlife Fund, the Center for International Forestry Research and financial support from Forest Governance Facility (FGF). The aim of these groups is to secure community rights in forests and promote community interests by influencing regional and national policy.

Another reason for the emergency of community networks is governance and political opportunity. In Guatemala, with the resolution of the long civil conflict, displaced citizens were returning to the Petén region and the government was under pressure to repatriate and resettle a large population. In Nepal, a multiparty parliamentary regime was established in 1990 after three decades of political struggle against the autocratic Panchayat regime; the new democratic polity provided space for civil society organizations to flourish. In Bolivia, conflict over control of forest resources, coupled with the growth of the Brazil nut sector, increased local demands for a greater share of benefits. In response, the government issued decrees to recognize local rights and ease tension between the *barraqueros* and rural communities and a community network evolved.

Charismatic leadership of the networks was crucial in all cases. In Nepal, in the absence of elected local governments, local cadres attracted to social issues became involved in movements for forest rights. In the Petén, the leaders played crucial roles in bringing diverse community groups under a single umbrella and consolidating their movement for community concessions. In Bolivia, some small producers active in the peasant federations' struggle for land emerged as leaders in the formation of COINACAPA.

Financial and technical support from national and international institutions has become instrumental in these cases. International cooperation organizations (particularly the Ford Foundation) in the Petén and in Nepal appear to have made significant contributions in nurturing ACOFOP and FECOFUN respectively during their early phases. In Bolivia, the Italian NGO Associazione di Cooperazione Rurale in Africa e America Latina (ACRA) provided both technical and financial support to COINACAPA. Such support helped these networks enhance their capacity, increase interaction among stakeholders and consolidate their actions. External support has also helped these networks connect with wider regional and global alliances, such as the Coordinating Association of Indigenous and Community Agroforestry in Central America (ACICAFOC), Global Alliance of Community Forestry (GACF) and the Rights and Resources Initiative (RRI).

Box 6.3 Philippines: Community-based forest management federation

The National Community Based Forest Management People's Organization Federation of the Philippines (National CBFM-PO Federation), formed in 2004, is the largest organized group in the Philippines, comprising 14 regional federations, 71 provincial federations and 1691 peoples' organizations that claim to represent more than 20 million forest residents. It is the umbrella organization defending the rights of forest-dependent communities, but it also seeks to help members to become ecologically accountable, economically viable, politically strong and socially responsive.

Whereas FECOFUN and ACOFOP evolved primarily to defend access to forest resources, COINACAPA was established to engage in the market. However, with the consolidation of power the former two networks have begun enhancing livelihood benefits. FECOFUN's attempts to introduce Forest Stewardship Council certification and to support user groups in enterprise development by helping them connect with private entrepreneurs can be seen in this light. The cases show that collective action at the level of community networks goes beyond securing forest tenure to enhancing livelihood benefits by promoting quality, achieving economies of scale and increasing bargaining power.

Strategies and actions

Networks seem to have incorporated similar strategies to advance their agendas. We observe similar patterns in their mobilization of institutional resources, external support and policy and legal tools. Non-material resources such as legitimacy, cohesiveness and symbolic capital are also being effectively mobilized. We identify the following strategies.

Building institutional and technical capacity Community networks and collective action at the secondary level help strengthen the institutional and technical capacity of member organizations. The FECOFUN and COINACAPA networks emerged where grassroots collective action appeared inadequate. The networks in turn supported capacity building among their member groups, supported by outsiders in areas of organizational management, recordkeeping, legal awareness, preparation of management plans, enterprise development, monitoring and evaluation and other professional and technical skills. The groups began improving their performance and more clearly defining their roles and have gained confidence in their actions.

Besides serving as political watchdogs and pressure groups, the networks have enhanced the productivity of their resources by adding value. For example, by assisting cooperatives, seeking markets, providing help in gaining market access and delivering information, ACOFOP has enhanced livelihood benefits from forest management. Enhancing institutional capacity by training members to maintain quality and comply with organic standards, as well as forming village-level groups to police local practice, became a central focus for COINACAPA. It helps small producers maintain standards set by fair trade rules so that the whole group can retain access to benefits. FECOFUN also has begun to support its members in achieving certification, to enhance their position in the market.

Capacity building of community groups and their networks is also the agenda of donors and external agencies. Donors and NGOs often have instrumental interests in the networks, which are considered good vehicles for delivering development. In our cases, the governments, international agencies and domestic NGOs appear keen to work with these networks. Direct dealing with them reduces transaction costs in participatory development. However, it is not always beneficial for the networks themselves, which do not always desire what the donors and NGOs offer.

Influencing public discourse and increasing legitimacy Community networks can influence public discourse on environmental resources governance. Before the growth of these networks, state agencies promoted state-centric discourse and action and sought solutions through bureaucratic management. The nationalization of forests in Nepal and establishment of the Maya Biosphere Reserve in Guatemala are actions indicative of this perspective. With the growth of these community networks, a counter discourse developed that effectively challenged the monopoly of state management and offered community management as a viable option. Meanwhile, the networks proactively engaged with stakeholders and lobbied for participatory, community-based management. They capitalized on and contributed to a shifting global focus on participatory resources governance and increased their alliances with civil society organizations. Gradually, community-based management has become not only an accepted but in many cases a preferred option.

Changing balance of power in favour of communities Networks bring agency to the tenure reform process. State-led tenure reform processes originate at the central level and are implemented through the bureaucracy, often treating local communities as passive recipients of state policies. Community networks and other secondary-level organizations, however, become active agents. Once the local communities become engaged, they can demonstrate their entrepreneurship and influence the reform process by collective expression.

Networks gain power through their mass base. The constituent members of the networks discussed here have strong social bases across large regions and can mobilize thousands of people to a common cause. Historically, local-level collective action was not adequately appreciated or recognized. In most cases, isolated groups acting alone were too weak to defend their rights from either centralized, bureaucratic power or external threats, whether private companies or other communities. Transfer of formal rights is only the first step. Maintaining the integrity of community concessions or community forestry remains a challenge, given the constant attempts by external actors to usurp resources. This is particularly apparent in the Petén and Nepal, where private companies and even government agencies have attempted to take back resources and community rights. Networks have defended common interests against such attempts. Because of the sheer number of members and the networks' ability to get national attention and call on national and even international allies, state authorities and market actors have begun to recognize and respect their petitions.

Meanwhile, interactions among the authorities, market agencies and community actors have changed. Previously, the communities interacted bilaterally with state authorities or market actors. Today, new platforms and mechanisms allow communities to interact along with other stakeholders in diverse contexts. The platforms have expanded from national to international levels. With these expanding contexts and arenas, the traditionally unequal relations of power between the authorities and communities are beginning to crumble. New configurations of power have emerged. For example, when government officials and FECOFUN leaders sit together in an international workshop and applaud community forestry, their relations take on a different form despite the level of conflict at home. These encounters serve as alternative channels of communication and resolution. Similarly, in many public programmes, district forest officers and FECOFUN leaders are invited and given equal status, a pattern that can gradually equalize their power relations.

Influencing policy and regulatory frameworks The networks' expanding activities have helped to establish strong links between local communities and political leaders. Conventionally, forest bureaucrats are the *de facto* policy-makers, legitimized by the political system, and often undermine local livelihood interests by imposing strict exclusionary regimes. Community networks have been able to protest against such policies through civil disobedience and by nationally denouncing threats against their rights. FECOFUN, for example, has worked as a watchdog organization since the late 1990s, constantly resisting any regressive policy decisions that undermine community rights.

Networks have changed their strategy, however, from purely resisting government decisions to proactively engaging with formal policy processes. Initially there were two problems:

1 the established party, the state authority, seldom listened to the voices and concerns raised by local communities; and
2 the community leaders neither had trust in the system nor the capacity to constructively engage and contribute to the policy process.

With the evolution of community networks the situation has changed. Apart from many state agencies' recognition of multistakeholder process, the networks themselves have evolved as legitimate actors and begun to expand their role. They have been able to capitalize on strategic relations with political leaders who favour participation of the community networks in policy-making, building support through them. This has proved to be an effective strategy for increasing participation in the formal policy process.

This practice has been gradually institutionalized in recent years. External donors and many multilateral environmental agreements have promoted the idea that local communities should be consulted on environmental policies. For example, the Convention on Biological Diversity and the reducing emissions from deforestation and forest degradation (REDD) efforts demand that local communities and civil society organizations be adequately informed and consulted during national policy formation. Networks provide a convenient way to meet this requirement. Because they represent both the local community and civil society, they have become crucial actors in all policy forums concerning forest and natural resources governance. FECOFUN has become a permanent member of the Forest Sector Coordination Committee and the District Forest

Coordination Committee in Nepal, for example, and ACOFOP is a civil society representative on the board of directors of the National Council of Protected Areas in Guatemala.

Outcomes of network building

Community networks have helped strengthen tenure in two ways:

1. enhancing collective action at the group level; and
2. increasing the bargaining power of the groups with government officials or with the market (as in the COINACAPA cooperative).

Through regular interaction and educational activities, the networks have helped develop the capacity of community groups. They have helped expand the interface with government agencies, markets and other civil society actors through formal multistakeholder dialogue and informal processes. At the broader level, they tend to promote democratic, inclusive, equitable and participatory discourses, even if these ideals are not always reflected in practice. For example, FECOFUN's stated agenda is to achieve 50 per cent representation of women in the community forest user groups (but has not yet met this goal).

State-led reform processes have created a favourable environment for enhancing people's access to resources, but many reforms do not automatically translate into increased livelihood benefits (Cronkleton et al, 2008; Paudel et al, 2008b), either because devolution is limited to subsistence use (e.g. in Nepal) or because it is not linked with the complementary services needed by the communities (in Latin America). Moreover, the state tends to invent new interventions hindering the reform or at least limiting the potential benefits that communities can draw from the reform process. In such contexts, initiatives by forest-dependent communities and their networks have helped modify state-led processes.

Several studies have observed that these networks have played important roles in enhancing access to forest resources and markets for forest products and securing rights over these resources (Plant and Hvalkof, 2001; Cronkleton et al, 2008). Similarly, Komarudin et al (2008), based on action research in Indonesia, conclude that collective action enhances tenure security and livelihood benefits of forest-dependent communities. They noted that although local-level collective action has minimized elite capture, higher-level network building and networking are necessary to increase access to land, raise incomes and improve women's status.

Challenges

Huge challenges remain. First, community networks must keep pace with the changing context to meet the expectations of their members and stakeholders. Second, the networks must help their members build the capacity to meet the standards and criteria set by governments and markets. Third, they must address governance issues, particularly equity and inclusion within the organization. Fourth, they must achieve a balance in their relationships with members, donors, government agencies, civil society organizations and media, particularly with their increasing involvement in international forums. As their membership and political influence grow, the networks come under the sharp scrutiny of donor agencies and NGOs. Finally, the experience of FECOFUN and ACOFOP shows that one-off policy reform is not enough; continuing vigilance against regressive policies or rollback of rights is required.

Important lessons can be drawn from the discussion on collective action, community networks and tenure security. Community networks emerged out of perceived crises in livelihoods and resource governance. These networks helped expand the scale and scope of collective action. Representing forest-dependent communities, networks have emerged as new actors in the management of forest commons. They play significant roles in translating progressive policies into practical realities so that communities can realize the livelihood benefits of tenure reform.

Notes

1. Between 1966 and 1996 Guatemala experienced a civil war. According to the Peace Accords, the socioeconomic and agrarian agreement establishes that at least 100,000 ha should be allocated to organized community groups.
2. Two important differences between community concessions and industrial concessions are determined by contracts. First, contracts give industries usufruct rights to manage timber products only, whereas community concessions are for integrated management, including both timber and non-timber resources. Second, whereas community concessions pay for the use of land (between US$1 and $1.50 per ha), industries pay the intrinsic value of timber depending on the species and the amount logged.
3. According to Ostrom (2000), exclusion rights include the collective right to determine who has rights of access, withdrawal and management and who does not (see Chapter 1).

Part IV
Regulations and Markets for Forest Products

7

Regulations as Barriers to Community Benefits in Tenure Reform

Juan M. Pulhin, Anne M. Larson and Pablo Pacheco[1]

Research increasingly indicates that strengthened forest tenure for communities and individuals can improve well-being, enable exclusion of outside claimants and improve forest management and conservation (Sunderlin et al, 2008). Despite such potential however, forest tenure reform remains tenuous and its impact limited. One reason is that even where substantial, secure rights have been granted, government regulations hinder community access to forest products and related markets. This chapter looks at the question of regulation; Chapter 8 addresses community engagement with markets more specifically.

Regulation refers to 'controlling human or societal behaviour by rules or restriction' (Koops et al, 2006, p81). Regulation can take many forms, ranging from formal legal restrictions promulgated by the government to less formal social regulation, such as norms that govern social behaviour in a given cultural context; this chapter refers primarily to the former. In forestry, 'regulations are rules prescribed to control the use of forest resources and to assure that the management of these resources conforms to government-defined standards' (Fay and Michon, 2003, p11). These rules are contained in state laws and their subordinate instruments – decrees, sub-decrees, policies, orders or circulars that constitute the 'regulatory framework' (Gilmour et al, 2005). Forestry rules are often enforced through the imposition of legal sanctions like imprisonment or fines as well as compliance with certain requirements, such as permits, leases, fees, management plans, monitoring and evaluation and other forms of regulatory instruments. With few exceptions, forestry regulation in the tropics is the responsibility of centralized government bureaucracies. Its stated objective may be maintaining the forest's economic and environmental services, though actual objectives may include less noble goals, such as maintaining government control of forestlands and forest resources. 'Permanent forest estates' in many tropical countries are often the legacy of colonial or European-inspired management approaches based on the exclusion of people (Fay and Michon, 2003) and 'double standards' in forest policy often prioritize logging companies over communities (Larson and Ribot, 2007).

Notwithstanding the recent efforts of governments and other institutions towards advancing forest tenure reform, accompanying regulatory frameworks have often limited the benefits for communities and individuals. Government rules, in terms of access, use and management of forestland and its resources, remain very strict (see Chapter 3) and their implementation overly bureaucratic. Such rules often limit the forests available for communities, restrict forest access and use and establish high transaction costs[2] that serve as barriers to the market, all of which limit the flow of forest benefits to local communities. They also promote regulatory capture[3] by powerful groups with strong economic interests and tend to breed corrupt practices in the forest bureaucracy. There is no evidence that regulatory frameworks as currently designed are the only or the best way to promote forest conservation.

Few researchers have systematically and comprehensively analysed regulations and transaction costs in the context of community forestry or communities living in forests.[4] Even more limited is literature directly related to forest regulations in the emerging forest tenure transition in many developing countries. This chapter seeks to answer several questions. What is the philosophical basis of strict government regulation in forestry, and how valid are its assumptions in the context of the emerging forest tenure transition? What types of forest regulations relate to forest tenure reform, and how do they promote the persistence of government control over management and use rights of community forests? How do the different regulations and transaction costs serve as barriers to markets and the flow of benefits to local communities? What theoretical insights and practical lessons can be distilled from the case studies, and what strategic actions can communities and other stakeholders take to promote a more responsive forest regulatory framework that will achieve the potential of tenure reforms?

The next section of this chapter traces the origin and philosophical basis of government regulations in forestry through a brief examination of the European tradition of 'scientific' forest management which persists in many tropical countries today. Drawing on the CIFOR-RRI case studies from Asia, Africa and Latin America, with supplements from relevant cases elsewhere, the third section explores different forms of forest regulations and how they undermine forest tenure reform efforts. Next, the discussion section synthesizes the findings and recommends strategies to contribute to a more effective and appropriate regulatory framework. This is followed by a short conclusion.

Origin and philosophical basis of government forestry regulations

The establishment of 'permanent forest estates' in many tropical countries and the strict regulations governing these areas are often the legacy of colonial or European forest management approaches that may be linked to the concept of 'territorialization'. Vandergeest and Peluso (1995, pp387–388) explain:

> *All modern states divide their territories into complex and overlapping political and economic zones, rearrange people and resources within these units, and* create regulations delineating how and by whom these areas can be used...*Territorialization is about excluding or including people within particular geographic boundaries and about controlling what people do and their access to natural resources within this boundary.* [authors' emphasis]

Territorialization as applied to forest estates has an ancient origin. The first clear record may come from Assyria, where in 700 BCE game reserves were set aside by decree for royal hunts (Dixon and Sherman, 1991). In medieval Europe, forests were demarcated as a particular domain in the *silva* (literally, a place for growing trees), reserved for the hunting pleasure of the dominant classes of landlords, namely the vassals of the sovereign (the nobles) and the monasteries (the clergy) (Fay and Michon, 2003). Most of the *silva* was owned by the monarch and the two dominant classes; the common people (the villeins and the serfs) were usually bound to a landlord and granted only restricted usufruct rights on the *silva* lands. As the population grew, tension increased between the common people, who needed farmland and forest resources, and the landlords, who wanted full and exclusive control of their forest domain.

Enforcement of forest regulations became the task of forest administrators with the specific mission of protecting the forest domain from encroachment. In Europe, the first royal corps of forest administrators (later called foresters) was created in 1290 to 'defend the royal rights of hunting and justice' and later to restrict the usufruct rights of peasants (Fay and Michon, 2003). This corps served the elite's economic interests. The increasing population, enclosure of land through privatization and loss of forestlands and their associated products, such as timber, fuel wood, fodder and game, caused the value of forests to soar. The kings and the nobles therefore used forest regulations not only to protect their exclusive hunting grounds but also to secure economic opportunities (Peluso, 1992). The establishment of forest estates was probably also based on calculations regarding the need for forest products and services over the long term.

Thus, forest regulations became the tool of the elite to restrict the exercise of usufruct rights, while a growing peasant population struggled to convert more lands for agriculture, expand grazing areas for livestock and acquire more firewood. Foresters and gamekeepers were employed to protect the political and economic interests of the royalty, nobility and clergy and exclude the common people from these areas. The tendency of foresters to exclude local people from the forests thus has a long history, dating back to the involvement of the forestry profession with landowning authorities. This attitude carried over easily into their involvement in the privatization of the commons in Europe, especially in the 18th and 19th centuries, and fit well with the undemocratic and hierarchical style of colonial authorities as well.

After a long period of repressive approaches, the administration of the forest domain became more constructive. Developing and harmonizing silvicultural practices to ensure sustained production became a major concern (Fay and Michon, 2003). In England, the application of a scientific revolution in forest conservation led to tree planting for economic purposes in the late 16th and early 17th centuries. Similarly, in 1661, Louis XIV of France and his minister of finance, Colbert, instituted revisions of forest administration and laws with the intent of reversing the reduction of forest cover caused by overexploitation (Elliott, 1996). From this time on, forestry embraced a more complex mission of regulating, administering, conserving and managing the forest domain.

The development of 'scientific forestry' from about 1765 to 1800, largely in Prussia and Saxony, provided legitimacy for territorialization and hence the enforcement of forest regulations to 'rationalize' forest management. Its emergence is best understood within the context of centralized state-making initiatives of the period (Scott, 1995). The early concept of scientific forestry was best captured by Le Roy, the warden of the park of Versailles, in Diderot's *Encyclopédie* of 1766:

> *In all ages, one has sensed the importance of preserving forests; they have always been regarded as the property of the state and administered in its name: religion itself had consecrated forests, doubtless to protect, through veneration, that which had to be conserved for public interests...Our oaks no longer proffer oracles...we must replace this cult by care, and whatever advantage one may previously have found in the respect that one had for forests, one can expect even more success from vigilance and economy...If one exploits wood for the present needs, one must also conserve them and plan for the future generations...It is therefore necessary that those who are charged with overseeing the maintenance of forests by the state be very experienced...they must know the workings of nature.* (Le Roy, cited in Harrison, 1992)

The above quote suggests the philosophical bases for designing and enforcing forest regulations by the state. First, forests are the property of the state and have to be administered in its name for the public interest. Therefore, a state forestry agency needs to be established to control forestlands and forest resources for the public good through regulations. Second, forests may be exploited to satisfy present needs but also have to be conserved for future generations. Thus, as the landlord, the state forest agency is both a forest enterprise and a conservation institution, roles that may be in conflict with each other (Peluso, 1992). Hence, forest regulations are needed to balance the economic and conservation objectives of state forest management. Third, those who are charged with overseeing the maintenance of forests by the state must be experienced and know the 'workings of nature'. This legitimized the mission of foresters and established the exclusivity of professional foresters in forest administration and management (Fay and Michon, 2003). As professional foresters discharge their functions, their actions, conducted in the name of the 'public interest', are guided and legitimated by forest regulations.

The first university training programme to promote scientific forest management was established at the University of Freiburg, followed by other universities in the German states in the 19th century. In 1824, a national school of forestry was founded in Nancy, France (Mantel, 1964) and it attracted students from all over Europe and the United States (Peluso, 1992). Forest science was based on technical calculations to achieve 'sustained yield' by applying silvicultural principles developed through experimental trials. When they returned home or travelled to colonies in Asia and Africa, or to Latin America, foresters carried with them the philosophy of state-controlled and technocratic forest management (Fernow, 1911).

The United States also played a role in influencing global forest management by shaping the forest conservation paradigm that continues to legitimize state management today, largely through Gifford Pinchot, who studied at Nancy and founded the US Forest Service. Considered the first proponent of 'modern resource conservation' (Eckersley, 1992), Pinchot believed in the complementarity of conservation and development: forests, he said, should be managed to 'provide the greatest good for the greatest number of people for the longest time' (Dana and Fairfax, 1980, p72). As a result, 'today, the term forest conservation can mean anything from intensive timber production to total preservation' (Elliott, 1996).

Both the utilitarian view of forests as a source of government revenue (forest use to provide the 'greatest good for the greatest number') and the more preservationist stance advocated by some conservationists have justified absolute state control of the forest resource base and the strict regulation of its use (see also Chapter 2). The ongoing delineation of large tracts of forestlands into production and protection areas by governments in many developing countries reflects the persistence of Pinchot's resource conservation paradigm. The same paradigm allows foresters to conduct their science according to the state's interests, even though they rarely view their policies or implementation as political acts. Today, scientific forestry refers to both the German tradition – regimented plantations with minimal diversity, and the foresters-know-best management for sustained yield – as well as the more modern concept of planned, sustainable, conservation-oriented professional management.[5]

Tribal peoples and other local communities gain little from state territorialization or nationalization of forest control except temporary employment as skilled or unskilled labourers on lands they probably once controlled (Peluso, 1992). Notwithstanding the promise that forest bureaucracies will manage forest resources wisely, their performance in many developing countries has been dismal, perpetuating or even exacerbating land degradation and rural poverty in many countries (Blaikie, 1985).

Despite recent efforts to provide new and secure rights to indigenous and other local communities through forest tenure reform, government regulations are still founded on the Euro-American scientific forest management tradition and the bureaucratic culture that has persisted in state forest agencies. As will be revealed in the following discussions of case studies from Asia, Latin America and Africa, forest regulations perpetuate state control over lands and forest resources, undermining the potential benefits of the reform.

Forestry regulations and tenure transition: Selected cases

Under new tenure arrangements in community forestry, forest regulations may be enforced through:

1. leases or classification systems that limit access to forestland, as in India;
2. conservation-inspired rules that limit activities in protected areas, as in Brazil; and
3. permits, agreements, taxes, management plans and similar requirements that limit access to timber and other valuable forest products, as in the Philippines, Guatemala and Nepal.

Regulations that limit access to land[6]

Despite the recent trend to devolve ownership and/or control of forests to communities, access to high-value forests may be restricted by zoning, classification systems and other land allocation regulations. Such regulations may be viewed as the state's first line of defence in securing valuable forestlands and limiting the area to be handed back to communities.[7] Such regulations may overlay all subsequent decisions, severely limiting community rights.

India is one example. As is typical of many Asian countries, India's forest management has a European legacy, in this case British colonial rule. India was one of the first nations to establish a professional forest service: it nationalized its forest domain under the Forest Act of 1865. Demarcation of uncultivated land under the management of the Indian Forest Service continued over the next century and throughout this period forests were valued mainly for their timber and contribution to the country's economic development. Tribal communities and other forest dwellers' resource rights were eroded as the state agencies and the private sector established greater control. Even after independence, much of the British colonial forest policy and administrative system continued to direct the governance of forestland and its resources (Poffenberger, 1996; Poffenberger et al, 1997).

To date, most of India's 77.47 million ha of forestland remains under state control. The country is endowed with rich forest resources containing about 8 per cent of global biodiversity, making it one of the 12 'mega-biodiversity' countries in the world. Yet more than a sixth of the country's geographical area (55.27 million ha) is considered 'wasteland'. This area has been the target of recent community-based forest management programmes, such as the Tree Growers' Cooperative Society (TGCS) programme, described in Chapter 3 and summarized briefly here.

The TGCS programme was a response to the growing concern in the 1980s over fuel wood and fodder scarcity and increasing land degradation. Its proponent, the National Wastelands Development Board, viewed the project as a more effective and sustainable institutional alternative for forestation than the existing Forest Department-led social forestry programme. A 'revenue village' is selected and formally registered as a cooperative.[8] The cooperative applies for a government lease in part of the 'revenue wasteland' (located in the village) that belongs to the Revenue Department of the state government. Such leases are usually for 25 years and can be renewed for the same term; they are one of the clearest cases of tenure transfer under community-based forest management in India. A study of TGCSs in the villages of Khoda Ganesh, Nathoothala and Kumhariya in Ajmer district of Rajasthan, however, demonstrates that forestry regulations actually perpetuate government control over forestland by limiting access to the more productive areas. Moreover, the Revenue Department retains the right to use the land for other purposes.

Although the villages' property rights have been temporarily secured under the new tenure arrangement, the livelihood benefits have been rather modest. Because of poor productivity of the plantations, the TGCSs have not been able to generate cash income, which was one of the goals. Even improvements in fuel wood and fodder availability – major goals of the programme – were not large. Only 28 per cent of 382 households surveyed reported increases in fuel wood availability, and 43 per cent reported increases in fodder. Also, the tree survival rate (43 per cent) was rather low, limiting the project's potential ecological benefits.

The reasons for the programme's limited impacts are multiple, but the state regulation that limits community access to productive forestland is a major factor. Each lease involves less than 40ha of land, irrespective of the population. Such small parcels are not enough to generate livelihood benefits for every household. Most leased lands are of poor quality and were highly degraded when they were handed over, necessitating difficult and costly development and requiring a long time to become productive. Indeed, about 59 per cent of the households surveyed in all three villages considered TGCSs 'unimportant' to their livelihoods.

In contrast, the government appears to gain from the new tenure arrangement in at least two ways. First, the TGCSs largely prevented encroachments on the leased sites and hence these areas have been protected. The TGCSs were therefore instrumental in preserving the village common lands, which legally belong to the Revenue Department. Indeed, encroachments and resource destruction have been noted in nearby areas not covered by lease arrangements. Second, the TGCSs helped improve the biophysical condition of the sites. Both these accomplishments further the conservation objectives of the state.

Limiting access to valuable forestland through forest regulations is of course not unique to India. For instance, Nepal, despite being among the pioneers of community forestry in Asia, has its own share of challenges in making productive forestland available to groups of forest users. In the *terai* region, where most of the productive forests are located, the Department of Forests retains greater control of high-value forests and has only rarely, and after grassroots demand, handed them over to community forest user groups (CFUGs) (Bhattarai, 2006; Ojha et al, 2008). As of 2005, only about 2 per cent of the *terai* forests had been handed over to CFUGs, compared with almost 24 per cent of the lower-quality hill forests. The government contends that products from these forests need to be distributed throughout the country, including to urban populations, and it should therefore be responsible for these areas. Indeed, the Forest Policy of 2000 imposed a 40 per cent tax on revenues generated from the sale of timber on the CFUGs in the *terai* and stipulated additional restrictions on forest devolution in this area (Bhattarai, 2006).

Similarly, in the Philippines, another country noted for its 'radical' and 'progressive' community forestry policy (Utting, 2000; Pulhin et al, 2007), communities continue to struggle to gain control over productive forest areas. Earlier government initiatives under the Integrated Social Forestry Program had leased to communities only lands already denuded of trees, then extracted cheap labour for reforestation and protection (McDermott, 2001). In addition, the government, through the Department of Environment and Natural Resources, has expected these communities to stabilize upland encroachment, increase the productivity of upland agriculture and control potential dissent. At the same time, the department retains the power to allocate timber concessions[9] (now called industrial forest management agreements) on residual forestlands when it is profitable and politically expedient to do so (Li, 2002). Although the transfer of forest management from the department to local communities over the past 25 years has been significant, with close to 4.7 million ha under various forms of land tenure instruments, the more productive areas in general are still under the remaining private timber concessions and agreements or under the government-controlled National Integrated Protected Area System.

As in India, the state-controlled forest management approaches of both Nepal and the Philippines are of colonial origin. Nepal's forest policies were directly influenced by the British, when its experts helped the Rana rulers establish the Department of Forests in 1942 (Paudel et al, 2008a). The department started the nationalization of forestland and perpetuated the colonial notion of scientific forestry in the country. The Philippines' forest management was a legacy of the Spanish and the American systems. The Spanish colonial government established the first forestry bureau, the *Inspección General de Montes*, in 1863, and introduced the European tradition of centralized forest management. The American colonizers who took over in 1898 then established a forestry school, in 1910, with the help of none other than Pinchot himself. The concepts of scientific forestry remain the basis for the country's forest resources management.

The use of forest regulations to limit communities' access to forestland is of course not solely an Asian phenomenon. In Cameroon, in Africa, the recent tenure arrangement that entitles communities to new bundles of rights to access, use and manage forest lands is not applied in the entire forest estate (Oyono et al, 2008; Diaw et al, 2008). A 1993 zoning plan classified the forestland into permanent and non-permanent forest estate. The permanent forest estate includes national parks, faunal reserves, game ranches, botanical gardens, zoological gardens, production forests (intended for timber extraction), protection forests and research forests – the richest, largest and most strategic forest areas. The non-permanent forest estate comprises less productive forests and agricultural lands adjacent to villages, and it is here that (at present) about 56 village communities have 25-year management agreements that entitle them to access, use and manage the land for livelihood purposes (Oyono et al, 2008). Hence, local communities have been legally excluded from high-value forests, which are largely reserved for commercial logging and for protected areas. Some community members argue that they have been given greater rights but to smaller areas, since they have customarily claimed access to a much larger area of forest (Oyono et al, 2008). The state's capacity to implement its regulations throughout the large forest areas it claims is also in question. As in Asia, the state's tendency to retain valuable forestlands in Cameroon is rooted in its colonial tradition (Oyono, 2004a).

Conservation and protection-oriented regulations

Conservation-inspired regulations – whether implemented by a forest management agency or a separate environmental agency – can also be used to limit the activities of communities that have been given legal rights in forest reserves and protected areas but have lost customary rights. Such regulations can limit use rights or require development and management plans that attempt to regulate the activities of local communities to achieve the state's conservation objectives. Porto de Moz, Brazil, is a case in point.

The first regulations seeking to control and monitor timber extraction and forest conversion in Brazil were issued in 1968 with the approval of the Forestry Code, the implementation of which was delegated to a federal environmental agency. In 1994, a decree for regulating forest management established mechanisms for ensuring reforestation and introduced forest management planning. Timber extractors already had to comply with similar procedures. In the late 1990s, environmental concerns prompted actions to demarcate conservation areas, which influenced the establishment of a national system of conservation units (*Sistema Nacional de Unidades de Conservação*) in 2000. To protect the rights of agro-extractive and traditional populations, 'extractive reserves' (RESEX) were created. Though previous land projects were under the jurisdiction of the National Institute for Colonization and Agrarian Reform, a RESEX was included as a specific type of conservation land use under Law No. 9.985, falling under environment agency jurisdiction.

In the Brazilian municipality of Porto de Moz, in the state of Pará, local communities have a history of struggles with timber and fishing companies. These companies used local resources, but communities reaped little benefits (Moreira and Hébette, 2003; Salgado and Kaimowitz, 2003). To protect their land and natural resources, communities demanded a RESEX. The resulting 'Verde para Sempre',[10] covering some 1.3 million ha and including about 58 communities, was created in 2004 by presidential decree. Although the reserve secured the property rights of residents and allowed the communities to exclude timber companies from their lands, it also imposed new constraints on forest use for smallholders living in the reserve.

The RESEX recognized the territorial rights of a mix of local communities and medium-scale landholders on the west bank of the Xingu River. Local people – influenced by NGOs and conservation organizations – adopted the RESEX model to formalize their land tenure rights as an expedient way to gain rights to an extensive area. The limiting factor is that the RESEX is a conservation unit. Landholders living inside it receive not full ownership rights but an indefinite usufruct right (*concessão de direito real de uso*), bounded by a variety of land-use constraints. In the RESEX, according to the law, the use of species at risk of extinction, practices that erode these species' habitats and practices that could harm the regeneration of natural ecosystems are prohibited. Timber extraction is allowed only when practiced in a sustainable way and only under special circumstances (e.g. when it is complementary to other extractive activities). Forest conversion is limited to 10 per cent of the total area, according to the rules established in the RESEX management plan. Also, the rules constrain the movement of water buffalos.

A RESEX is intended as an area where landholders develop extractive activities and small-scale agriculture. Hence there are no limits on the collection of non-timber forest products (NTFPs), but other consumptive uses, such as logging (when allowed at all), require a forest management plan. Any activity to be developed in the RESEX must be part of a RESEX development plan. The communities cannot use timber resources, for example, until they have completed such a plan, which can be undertaken only after the definitive development plan for the whole RESEX has been written and approved – which has still not occurred.

Hence four communities that demarcated their lands with the assistance of a forestry project have not been able to develop forest management plans. Exceptions were made, however, for two such communities because they were supported before the creation of the RESEX by the ProManejo programme, a federal project that supports community forestry through the development of low-intensity harvesting and artisanal wood transformation projects.

A distinction is made between high- and low-intensity plans, but both are subject to the same bureaucratic steps. The low-intensity plans are somewhat simpler, but all plans must be signed by a professional forester and in community areas plans also have to be signed by leaders representing the community or territory. The professional forester, who helps write the plan, is at the same time responsible for the forestry operations in the area. In theory, this system should ensure relative transparency in the formulation and implementation of forest management plans, facilitating central agency supervision of the plan's implementation.

Sometimes timber is harvested even if a community does not yet have a formal management plan. Networks of sawyers, local loggers, traders and truckers who were employed by the timber companies prior to the creation of the RESEX have been taken over by local politicians, who use their influence and connections to extract timber and supply timber industries in the city, in contravention of the regulations. The volume of these informal transactions cannot be estimated, though there is said to be less logging in the reserve than before.

The Porto de Moz case demonstrates that despite grassroots mobilization to create a reserve, the government's environmental and conservation objectives tend to dominate the interests of the local population. The formal institutions are highly bureaucratic and ineffective in implementing their own regulations. The lack of a management plan prevented some local communities from pursuing commercial logging operations, even though the system of extractive reserves was intended to protect the interests of agro-extractivist communities and people whose traditional livelihoods depend on timber and non-timber forest products. The conservation-oriented regulations leave local people little flexibility to use the resources to fulfil their material needs – at least not legally. Neighbouring communities are now seeking other models for their land claims.

Regulations that limit commercial use of valuable resources

State forest agencies sometimes act as forest enterprise organizations that regulate the commercial use of valuable forest resources, such as timber, in the name of the public interest. Even where valuable forest resources have been handed over to communities by the state under the new tenure arrangements, strict government regulations still constrain the flow of benefits to local communities. Three cases illustrate this issue.

Ngan Panansalan Pagsabangan Forest Resources Development Cooperative, the Philippines This cooperative is one of 1781 People's Organizations in the Philippines involved in the government's community-based forest management (CBFM) programme. Located in the Compostela Valley Province on the island of Mindanao, it manages 14,800ha under a 25-year tenure instrument that entitles the cooperative to manage and utilize the timber resources in accordance with the principles of sustainable forest management (see also Chapter 3).

The cooperative was one of the earliest government experiments in facilitating a transition from a corporate timber enterprise to a community-based approach to achieve the goals of sustainable forestry and social justice. Although the timber enterprise is managed by technically competent professionals (mostly former employees of the logging company that operated in the area from 1969 to 1994), major policy decisions rest with the cooperative's general assembly and board of directors, with representation from the Mandaya-Mansaka tribal group. This organizational setup allows the cooperative to function as a business entity. It is the only community forest in the country and the first in Southeast Asia to be certified by SmartWood, having met the criteria for a sustainably managed forest in 2000.

The cooperative was established in 1996 and given the rights and responsibilities to manage and protect the assigned forestland in the towns of Compostela and New Bataan in Compostela Valley Province. The overall management of the area is governed by the 'community resource management framework'. Produced by the cooperative with the support of the Department of Environment and Natural Resources, this document serves as the basis for forest management activities and strategies for 25 years. The cooperative also had to prepare a five-year work plan to guide its operations. The resulting work plan indicates how much of the production forest will be subject to harvesting during the first five years and how much timber will be extracted from the forest plantation (Pulhin and Ramírez, 2006). The earnings will fund forest development and protection projects, such as reforestation, agroforestry, timber stand improvement, assisted natural regeneration, foot patrols and the establishment of checkpoints to ensure continuous forest protection. It will also generate livelihood initiatives for the cooperative. The development of additional tree plantations and agroforestry areas is expected to reduce the pressure on natural forests as a main source of timber and improve forest cover, minimize erosion and the occurrence of flash floods.

In the Philippines, the preparation of comprehensive management plans, such as the management framework and the five-year work plan, is often beyond the capabilities of People's Organizations. Professional foresters must often be hired – something they typically cannot afford, or that at least increases their costs. Although this particular cooperative can prepare its own plans, thanks to the extensive technical experience of some members, the process nonetheless involves a lot of time, effort, negotiation skills and transaction costs, from timber inventory to approval.

Although the approved work plan specifies the target volume to be harvested annually, the actual volume harvested depends on approval from the Department of Environment and Natural Resources, which issues an annual permit. The real volume harvested is usually lower than the one proposed in the work plan. Without the permit, the cooperative cannot proceed with its timber harvesting operations, but approval can easily take more than six months, in part because it is issued by the department's central office in Metro Manila, leaving the cooperative with only six months to operate. Total costs can be as high as US$4,700 (see Figure 7.1).

Even after the permit has been issued and the timber has been cut, regulations to control the transport of harvested timber create additional problems (Dugan and Pulhin, 2006). Communities must obtain a permit for moving timber to the roadside and another to transport the timber to buyers. Further delays and additional transaction costs ensue because the department staff who issue the permit are usually many kilometres away. In the Philippines, tree farmers who

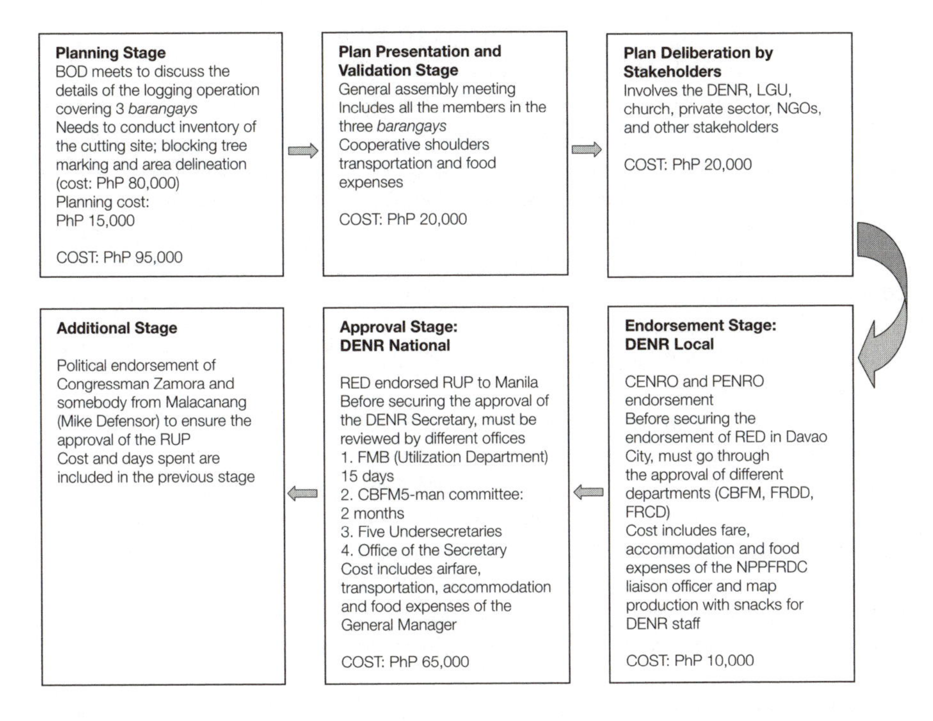

RUP, resource use permit; NPPFRDC, Ngan Panansalan Pagsabangan Forest Resources Development Cooperative; exchange rate, US$1 ≈ PhP44

Source: Puhlin et al (2008)

Figure 7.1 *Application process and transaction costs for Ngan Panansalan Pagsabangan Forest Resources Development Cooperative, 2006–2007*

develop plantations on their own private lands have also complained bitterly about these transport permits, which were originally intended to monitor and control the removal of timber from natural forests. The complexity of regulations and procedures has fuelled corruption, since each step creates the opportunity to extract money from communities (Dugan and Pulhin, 2006).

Community forest concessions, Petén, Guatemala[11] The community forest concessions in the Petén are located in the multiple-use zone of the Maya Biosphere Reserve. The reserve, consisting of 2.1 million ha, was established in 1990 as a conservation area to protect natural and cultural resources and the associated goods and services. Faced with pressure from community organizations and conservationists, the Guatemalan National Council for Protected Areas (*Consejo Nacional de Areas Protegidas*) saw the communities already living in the reserve as a potential ally to facilitate their work and there are currently 12 community concessions inside the reserve covering a total area of 426,000ha (see also Chapter 3). All but one had achieved Forest Stewardship Council (FSC) certification by 2006.

The contracts grant the community concession holders an exclusive use right over the defined area and its resources for 25 years. Unlike the extractive reserve in Porto de Moz, Brazil, which limits the commercial use of timber in the name of forest conservation, the Petén community concessions allow the use of valuable timber and non-timber forest products, although under strict regulations. The communities applying for concessions were required to incorporate as legal entities with formal bylaws and internal regulations to take legal responsibility for their concession. The organizations were initially required to sign technical assistance contracts with local NGOs, although this is no longer the case.

To regulate resource extraction, all concessionaires are required to develop sustainable management plans for each product harvested, including NTFPs, to be approved by the Council for Protected Areas. Timber management plans include full inventories of resources, environmental impact assessments and detailed plans for harvesting operations. Annual operation plans must also be developed and approved. Concessionaires are required to file operational reports every semester, pay various taxes and fees and acquire FSC certification. Failure to meet these rules and responsibilities could mean cancellation of the concession contract.

Those regulations are a substantial burden. The startup costs are difficult to calculate but are probably high, considering that NGOs and projects usually make substantial investments in training and equipment. Somewhere near US$10 million was directed at creating the startup conditions for community forest enterprises and introducing and subsidizing the certification scheme. The direct costs of creating the initial organizations are estimated at US$2,000 each. Preparation and approval of annual operating plans account for 5 per cent to 8 per cent of operating costs.

The regulations seen in the Petén concessions appear fairly typical of community forestry and of forestry permits in general. For instance, a study by Navarro et al (2007) in Honduras found that obtaining a logging permit involved 20 actors, 53 procedures and 71 steps and took an average of three to four months. Similarly, in Costa Rica, the process involved 11 actors, 31 procedures and 34 steps, and could take up to 18 months. A related study in Nicaragua's autonomous regions identified around 30 steps for areas over 500 ha (Navarro et al, 2008). In the Bolivian site Cururú, it took longer than two years from the initiation of the management plan to its final approval.

In Nicaragua, the costs of the general management plan and the environmental impact assessment for the management area were about US$2 and US$1 per hectare, respectively (Argüello, 2008). Annual operating plans covering the annual extraction area range from US$9–12 per hectare for broadleaf forests. The initial investment for these studies at Layasiksa, one of the CIFOR-RRI study sites, was more than US$50,000 because the area covered extended beyond the parcels managed for logging. The process for establishing community forests is so complicated in Cameroon that no community has been able to establish a community forest without extensive external assistance (Oyono, 2002, 2004b); the required management plan can cost as much as US$55,000 and take up to two years to complete (Smith, 2006). In addition, logging must be undertaken using low-impact procedures. In contrast, short-term concessions to the private sector, known as *ventes de coupe*, are less regulated, entailing no management plan and no restrictions on logging methods (Oyono et al, 2006).[12]

The combination of complex bureaucracies, high upfront costs in time and money, the lack of credit and the risk associated with demanding formal markets present major disincentives for community investment in formal management plans. Under such conditions it is very unlikely that communities will undertake community-based operations without significant outside support or other incentives. Indeed, the Petén community forest enterprise model is unlikely to have been successful without the infusion of high external support during its initiation period.

It should be mentioned, however, that some governments have also provided technical and financial assistance, grants and subsidies to communities, although these are usually quite inadequate. In Mexico, for instance, the state requires that communities have forest management plans but also provides funds for this.[13] By and large, however, as demonstrated in many countries, government support is usually inadequate if not absent, and thus significant external support is needed in the formulation of formal management plans and other bureaucratic requirements.

Nepal community forest user groups (CFUGs)[14] Nepal's regulations for community forests leave ample room for government foresters to interfere with the rights of user groups. Even after communities have satisfied the regulatory requirements, additional burdens hinder the marketing of these products and thus the flow of economic benefits to local communities.

Nepal's Forest Act of 1993, Forest Regulations of 1995 and community forestry guidelines of 1999 provide the legal basis for handing over patches of national forests to CFUGs and identify the roles and authority of the district forest officer and the CFUGs to ensure sustainable management. The main contractual document that guides forest management practice is an operational plan, prepared and agreed upon by the district forest officer and the CFUG.

Along with changing livelihood strategies, new market opportunities for forest products and services have been emerging recently. There is a well-established market for timber and the market value for some NTFPs has increased in recent years. Despite this, market opportunities associated with high-value timber and NTFPs have not been fully utilized. Excessive regulations and associated transaction costs are major barriers preventing the flow of economic benefits.

Although the transfer of national forests to CFUGs involves the right to manage and sell valuable forest products, the actual benefits of this reform are constrained by a complicated system of approval for operational plans, annual harvests, sale of forest products outside the group and any necessary amendments in the user group's rules. District forest officers often use their administrative and technical influence to add provisions beyond what is legally required. For example, the operational plan of Sundari CFUG includes a provision stating that when harvesting timber from the community forest, the CFUG should get permission from the district forest officer and record the harvested amounts by species. A government circular of 1996 instructs CFUGs first to satisfy internal demand, then that of neighbouring CFUGs and adjacent districts before finally being permitted to sell their products in the open market (see Chapter 8). Bureaucratic hassles involved in timber trade are discouraging. In one case, a CFUG member who wanted to sell 300 cubic feet of excess timber in the market had to visit the range post more than 12 times over a four-month period before getting the final approval.

Such regulation also discourages outside buyers, who prefer to purchase timber from other sources, particularly government sources. This consequently reduces competition for CFUG timber, resulting in low prices. In addition, the imposition of a 15 per cent tax on the sale of certain species has created additional disincentives to sell timber outside the CFUG. Communities believe that anything with high commercial value unnecessarily draws authorities' attention and that they enjoy more autonomy if they manage resources only for subsistence purposes. One option is to enter into non-transparent transactions or even illegal activities to sell their forest products. This is a serious problem particularly in the NTFP trade. Producers and collectors are in a weak position with low bargaining power and they have no control over the long and non-transparent market chain; they become price-takers.[15] Another option is to resort to bribery, which is an easier way to persuade government officials than fulfilling difficult formal requirements. The saying that 'more regulation means more corruption' is well understood in the Nepalese bureaucracy in general and in forestry in particular (Paudel et al, 2006).

Discussion

The cases highlight three types of regulations that influence the outcomes of forest tenure reforms. The first type limits the area available to communities in terms of size and/or quality of forests. Examples are India's Tree Growers' Cooperative Society programme, Nepal's recent policy limiting access to the more productive areas of the *terai*, the Philippines' earlier government initiatives under the Integrated Social Forestry Program that allocated to communities only land already denuded of trees and Cameroon's policy limiting community forestry to the less productive nonpermanent forest estate. In all these cases, the states' tendency to retain valuable forestland is rooted in their colonial tradition and perpetuated by modern forest bureaucracies.

The second type of regulation emphasizes conservation by delineating conservation areas and imposing limits on use. This is illustrated by the extractive reserve in Porto de Moz, Brazil. Even though grassroots mobilization was what led to the reserve, the government's environmental and conservation objectives have tended to prevail over the interests of local populations.

The third type of regulation imposes bureaucratic requirements that restrict communities' commercial use of valuable forest resources. The experiences of the forest resources development cooperative in the Philippines, the community concessions in the Petén, Guatemala, and the CFUGs in Nepal demonstrate the complex processes and high transaction costs involved in the commercial use and marketing of valuable timber and NTFPs, sometimes even after communities have satisfied regulatory requirements.[16]

The regulatory frameworks accompanying tenure reform have often limited the benefits to communities and individuals. The first type of regulation restricts the potential contribution of more valuable resources to livelihoods and poverty alleviation. Overemphasis on conservation objectives and related limits on resource use may place unreasonable (as well as unenforceable) limits on livelihoods. Imposing excessive bureaucratic requirements for commercial use and marketing of valuable forest products makes it almost impossible for communities to participate in formal markets without outside support or losing resource rents through elite capture. This in turn minimizes the flow of economic benefits to local communities, reducing the potential of the tenure reform to advance livelihoods and alleviate poverty in rural areas (see Chapter 8).

That regulations accompanying forest tenure reforms have negative effects on communities is clear, but it is more difficult to isolate impacts of forest regulations on forest conditions. Some regulations are not really enforced, as the cases from Brazil and Nepal suggest. Nevertheless, trends in forest conditions are generally positive in the different cases analysed, except in Cameroon, where conditions have declined. India's Tree Growers' Cooperative Society model, despite limited area coverage and less productive lands, has improved forest conditions. The forests in Brazil's extractive reserve programme may have marginally improved with the exclusion of the larger loggers, although some regulations are not actually enforced. Conditions in the Ngan Panansalan Pagsabangan forest have slightly improved, conditions in the Petén are substantially better than in surrounding non-managed areas and in Nepal significant improvements in forest condition were noted (see Chapter 9). Such improvements, however, may be attributed to other factors, such as the nature of the reform, its associated local and external support and/or the location of the forest, besides regulations. Indeed, in some cases, such as in Cameroon, tight regulations may have worked against both people and forests, causing declines in forest conditions and little improvement in livelihoods.

States may not necessarily have sinister motives in controlling forestland and its resources. Forestry agencies have justifications for all three types of regulations, which may be based on economic, moral and technical grounds. Limiting the area for communities may be aimed at getting the highest possible rents for the state in the most efficient way, though it may also involve private gain or doing political favours for certain supporters.[17] Regulations to limit use of conservation areas are founded on the idea that forests should serve the public interest and hence have a moral logic. The third type of regulation – bureaucratic requirements – is associated with forestry's culture, which holds that foresters possess superior technical knowledge and hence need to control all forest operations to ensure a more rational form of forest management and use. At the same time, regulations may represent growing and sometimes self-serving bureaucratic systems, untouched by reforms intended to reduce burdens, corruption and inequities.[18]

Whatever their rationale, regulatory frameworks tend to undermine the goal of local appropriation and the expectation that greater local 'ownership' (literally and figuratively) or 'buy-in' will lead to long-term commitment, income and hence sustainability in forests. Excessive regulations, in particular, interfere with such ownership, may override and weaken effective traditional practices and could encourage profitable illegal activity. This is demonstrated in Brazil, where excessive regulations on land use and forest management result in non-compliance and 'illegal' timber harvesting by the local communities in their extractive reserve.

This is not to imply that all state forestry regulation is unnecessary. There is no question that clear and enforceable rules can protect and improve forest conditions – and that the state often has a role to play in both rule-making and enforcement. What is in question, however, is what kind of regulation and how much is needed to achieve outcomes that balance the objectives of improving both livelihoods and forest condition. Obviously, regulations that are enforceable are more useful. Sometimes, rather than improving unworkable regulations, and hence improving quality, governments simply issue *more* regulations. This is what happened in 2006 in Nicaragua. There, the failure to enforce the forestry law led to a forest emergency declaration and then a moratorium prohibiting timber exports and the logging of certain species and establishing no-logging zones. In addition, more entities were included in forest law enforcement – which only resulted in more chaos, at least for a time.

How, then, could forest regulations better serve the interests of local communities and promote forest sustainability under new tenure arrangements? Lessons learned in many countries indicate general principles for successful implementation of community forestry initiatives (Gilmour et al, 2005):

- avoiding over-regulation so that the partners in implementation, particularly the local communities, can comply;
- starting with simple initiatives and adding complexity based on the ability of partners to handle increasingly complex tasks;
- minimizing transaction costs for all partners.

These principles of course are easier said than done. Recommendations like 'removing regulatory barriers' (Scherr et al, 2003) and 'deregulation' (Fay and Michon, 2003) seem not to appreciate that these are issues of power relations rather than administrative or technical concerns. It would be naive to assume that forestry agencies will easily relinquish regulatory power and give local communities more control over forest access, use and management under the new tenure arrangements. Even in countries such as Nepal and the Philippines, which started with progressive community forestry policies, new sets of regulations or policies can easily undermine earlier initiatives. In Nepal, for instance, the progressive Forest Act of 1993 and Regulation of 1995, which granted greater rights to CFUGs to manage community forests and promoted more equitable sharing of forest benefits, have been diluted by recent decisions: the Forest Policy of 2000 and the 2003 Collaborative Forest Management Plan undercut previous rights and imposed a 40 per cent tax on revenue generated from the sale of timber in the *terai*. Similarly, in the Philippines, the early momentum and optimism associated with adopting community-based forest management was dampened by a series of national permit suspensions and the attempt to cancel all community forest agreements except those with foreign funding.

Attempts to simplify regulations have not been encouraging in many parts of the world. In Brazil, although simplified plans are easier to develop, obtaining approval is still difficult (Carvalheiro, 2008). In Nicaragua, simplified plans were developed to salvage timber affected by Hurricane Felix in September 2007, but six months later, as the wood rotted and the rainy season approached, communities were still awaiting formal approval (Larson et al, 2008). Even these plans required the signature of a forester and hence entailed a financial investment. Exactly the same situation has been observed in the Philippines. Securing salvage permits to sell trees felled by typhoons, even if the trees had been planted by the farmers themselves in their community forest areas, can take more than six months and involve many transaction costs.

Enabling regulatory frameworks cannot be developed overnight. They are often a product of long and continuing struggles by strong community alliances that must be able to wield countervailing power to challenge the territorializing behaviour of the state (see Chapter 6). In Nepal, the imposition of the 40 per cent tax on timber revenue was challenged by the Federation of Community Forest Users, Nepal (FECOFUN)[19] in the Supreme Court (Bhattarai, 2006). The court declared the regulation unconstitutional and the government eventually reduced the tax to 15 per cent. FECOFUN has successfully tackled other national and local issues confronting community forestry in Nepal (Paudel et al, 2008a; Bhattarai, 2006). In Guatemala, the Association of Forest Communities of Petén went to court over a proposed government project in 2003 aimed at expanding the protected area around the Mirador Basin by dissolving the community forest concessions and integrating community members into private 'sustainable ecotourism initiatives'. The association argued that the affected communities had not been consulted and that the plan could actually increase pressure on the reserve (ACOFOP, 2005; Gómez and Méndez, 2005). Guatemala's Supreme Court ruled in its favour and declared the project illegal in mid-2005. As such experiences demonstrate, investments in building strong community alliances constitute a key strategy for making the existing forest regulatory framework more responsive.

Considering the tendency of forestry agencies – with their bureaucratic traditions and regulatory mandates – to craft and enforce strict regulations, community networks and their allies should advocate for simple and enforceable regulations that build on existing rights and management practices (Larson and Ribot, 2007). To enhance credibility and generate strong external support, community alliances such as FECOFUN should also police their own ranks, ensure accountability mechanisms and encourage cooperative values among members.

More importantly, higher levels of government, NGOs, donors and the grassroots all need to effect a paradigm shift on the discourse regarding local people, forests, conservation and sustainable use of forests. In many developing regions, there is still a strong dichotomy between protection and destruction (read: logging) and forest agencies still treat local people as a resource to provide cheap labour in forest rehabilitation or an instrument to achieve biodiversity and related conservation objectives. Such a discourse needs to be replaced with a rights-based philosophy of forest management (Larson and Ribot, 2007) that grants greater sovereignty to local communities over forest resources without sacrificing sustainability. Such a paradigm needs to influence national forestry schools and shape the next generation of foresters, with the goal of institutionalizing new ideas about forests and regulations that build on local strengths and capacities and helping foresters become facilitators rather than purely regulatory agents. In the medium term, the paradigm also calls for the reinvention of forestry agencies, which can devolve not only responsibilities but also authority to local communities, change outmoded regulatory policies and procedures and retool staff with skills in negotiation, conflict resolution and extension service to better serve local communities.

A major challenge is how a state regulatory framework can accommodate diverse local realities (including self-regulation) in a way that improves local livelihoods and alleviates poverty without undermining the productive, environmental, cultural and other values that forests provide. With new schemes such as REDD (reducing emissions from deforestation and forest degradation) likely to complicate regulations even further, it is crucial to clarify rights and tenure to protect communities from the excesses of the state's regulating power.

Notes

1. We wish to thank Sushil Saigal and Phil René Oyono for their inputs on the India and Cameroon cases, respectively.
2. In this chapter, transaction costs refer to the costs (financial and other) associated with complying with regulations and bureaucratic requirements, such as obtaining leases, agreements or permits and preparing management plans.
3. Regulatory capture refers to situations in which a government regulatory agency, such as the forest department created to act in the public interest, instead acts in favour of the commercial or special interests of parties other than local communities.
4. For examples see O'Brien et al (2005) and Adhikari and Lovett (2006). Verifor has compiled useful studies of 'forest verification systems' but without specific emphasis on communities (see www.verifor.org, last accessed September 2009).
5. We wish to thank Timothy Synnott for pointing out these two different definitions.

6. Except when other references are cited, this section is largely drawn from the CIFOR-RRI country report on India by Saigal et al (2008).
7. Thanks to Deborah Barry for this point.
8. A revenue village is an administrative category in India, which is below district and *tehsil*. It has defined boundaries and may contain one or more hamlets.
9. More recently, expiring timber concessions are being converted into industrial forest management agreements (IFMA) to meet legal requirements under the 1987 Philippine Constitution. However, some critics argue that IFMA is just a redressed version of the timber concession system.
10. In translation, 'Green Forever'.
11. Except when other references are cited, this section is largely drawn from Larson et al (2008a).
12. Compared with forestry, the agriculture and livestock sectors are less regulated and hence involve much lower transaction costs. In some cases, this creates perverse incentives to deforest, as people prefer to pursue agriculture and ranching in forestland.
13. This is not as positive as it might sound, however, since what it does in practice is guarantee payments to foresters even if communities develop the plans themselves.
14. Except when other references are cited, this section is largely drawn from Paudel et al (2008a).
15. A price-taker is an economic actor that must accept the prevailing market price for its products because its own transactions are unable to affect the price.
16. In many cases, larger private sector actors complain about the burden of forestry regulations as well. Still, in general, they are better equipped than communities to deal with regulations and bureaucracies; they also often receive special treatment because of their personal and political connections (see Larson and Ribot, 2007).
17. An example of doing political favours for certain supporters is well exemplified in the case of the Philippines during the Marcos administration, when the issuance of timber licence agreements in productive forests was used to gain political support from the elite group while communities were allocated denuded areas to reforest, thereby providing cheap labour to the government.
18. Thanks to Timothy Synnott for pointing this out.
19. FECOFUN is a national federation of forest users across Nepal with membership of almost 12,000 formally registered user groups dedicated to promoting and protecting users' rights (see Chapter 6).

8
Communities and Forest Markets: Assessing the Benefits from Diverse Forms of Engagement

Pablo Pacheco and Naya S. Paudel

This chapter explores the engagement of smallholders and communities with forest markets in the context of tenure reforms. Tenure reforms, in theory, should enhance these actors' access to forest resources and thus improve the benefits accruing from commercial forest resources use. In practice, however, it is not that simple. On the one hand, markets for forest products can provide alternative sources of income streams for smallholders and communities.[1] On the other hand, they may constitute a channel for the transfer of economic rents to other actors better positioned in forest markets. These two situations coexist to differing degrees. Most often, communities that choose to market their forest products, whether timber or non-timber forest products (NTFP), are able to generate cash income but often do not earn as much as expected because of benefit flows to traders, intermediaries and timber processors up the value chain and thus communities are often relegated to a role as raw material providers.

This chapter argues that the income that communities capture from the sale of forest resources depends not only on their ability to manage and process these resources effectively and efficiently, in both economic and ecological terms, but also on market factors. These factors are clustered into three sets of issues: the capacities that smallholders and communities have to interact in forest markets, the conditions under which such markets work and the ways in which communities engage in markets. A typology encompassing four kinds of community engagement with forest markets is developed, and seven cases from four countries provide examples of how these situations unfold in practice.

The next section provides a brief literature review about community forestry and market integration. We then explore three sets of variables (community capabilities, market development and forms of market engagement) to build a typology depicting four situations of community integration to forest markets. Case studies of communities under the different situations of market engagement follow. The subsequent section discusses ways to enhance forest benefits to communities. Finally, the last section concludes the chapter.

Relationships between communities and forest markets

According to Hayami (1998), two contradictory perspectives dominate the analysis of community interactions with markets. Some analysts see market engagement as an avenue for reversing the situation of deprivation that communities often face (Hallberg, 2000) and communities have to enhance their capacity to compete in the market. Others see markets as entities working against the poor, since they facilitate surplus appropriation and transfer from subordinated sectors, such as rural communities, to economic sectors, such as logging companies or industry (Watts and Goodman, 1997), and maintain that integration to markets is not a panacea for enhancing forest users' livelihoods (Pokorny and Johnson, 2008). The former perspective dominates the current discussion (see Donovan et al, 2008a, 2008b).

The relationships between communities and markets are determined in part by the capabilities that communities have to compete and to derive benefits from markets. A growing body of literature suggests that communities must improve their competitive position in such markets by creating and managing forestry enterprises, establishing long-term relationships with buyers, processing marketable products and gaining access to financial capital (Donovan et al, 2008b). Access to markets and to information about market conditions, along with bargaining power and negotiating expertise, are also important factors (Macqueen, 2008). Adding value is assumed to be desirable, and this implies vertical integration of production and processing (Donovan et al, 2006). In many cases, it is likely that vertical integration enhances competitiveness in the marketplace, in that more integrated enterprises become more competitive and thus obtain higher profits, but that may not always be true (Antinori, 2005).

Communities' ability to benefit from markets is also shaped by market structure. Molnar et al (2007) observe that local wood producers are increasingly forced to compete with low-cost, high-volume multinational companies. The growing importance of domestic markets, however, tends to work in favour of community forestry. These authors suggest that producers can find competitive advantage in lower transportation costs and enough supply flexibility to satisfy domestic market demands. Markets for NTFPs are highly diverse and accordingly require different livelihood strategies (Belcher et al, 2005): some are sold in local markets, others (for example Brazil nuts, Stoian, 2004) reach distant regional and international markets.

Yet another factor affecting the relationship between communities and markets is the regulatory framework. To participate in markets, smallholders and communities need a level playing field, without institutional hindrances, high transaction costs or direct barriers (Kaimowitz, 2003a). Conditions that favour small-scale producers and communities are low costs of market entry and a low-cost regulatory environment with minimal harvest, transport and sales permits (Scherr et al, 2004).

Some researchers have examined the relationships taking place outside the forestry regulations – the informal and illegal marketplace. The drivers of illegal logging have been explained elsewhere (Contreras-Hermosilla, 2001). Illegal logging has important implications for market functioning. It distorts timber markets by depressing prices, but many people, including the poor and unemployed, derive income from illegal logging and consumers may also benefit from the lower prices (Tacconi, 2007b). The environmental implications of illegal logging are ambiguous, although it is likely that those with more limited access to markets and less capital are going to destroy less forest than wealthier groups (Tacconi, 2007b). Informal markets tend to show more asymmetric relationships and often penalize the sellers with lower prices (Pacheco et al, 2008b).

This chapter does not consider markets either virtuous or evil, but depending on how they operate in practice they may benefit smallholders and communities, or they may disadvantage these actors by transferring economic rents from forest resources to others. Markets may even do both of these things at the same time. We argue here that whether markets help or harm communities depends on both endogenous factors (community capacities to compete) and exogenous factors (market conditions and policy environments).

Typology for assessing market engagement

An analytical framework helps us to understand community interactions with markets by looking at three sets of factors. The first are related to the *community's capabilities* for engaging in the markets – physical access to markets, bargaining power, knowledge of market dynamics and organizational capacities. The second are factors relating to conditions of *market development* – price distortions, incomplete information due to asymmetric relationships and state control over the market, among others. The third set of factors involves *forms of market engagement* – the type of product and whether the relationship with the market is stable or sporadic.

Seven case studies have been selected to explore how community capabilities, market development and forms of market engagement affect benefits from commercial forest resource use. Five of the seven cases come from Latin America, where communities are involved in timber markets to a greater degree than in other regions.

The first two cases engage indigenous communities managing their forests with commercial aims: Layasiksa in the North Atlantic Autonomous Region of Nicaragua and Cururú in Guarayos, Bolivia. Here, market relationships are mediated by a community forest enterprise, mainly for commercial logging, which manages activities along the value chain from logging to timber processing and commercialization. The next two cases are a community in the Bolivian northern Amazon and the Suspa community forestry user group in Dolkha,

Nepal. Both communities depend largely on the sale of NTFPs. The next two cases – smallholders in Iturralde, La Paz, Bolivia, and in Porto de Moz, Brazil – represent situations in which smallholders make individual decisions over their forests, often outside the law. Finally, the Sundari community forest user group (CFUG), in Nawalparasi, in the *terai* of Nepal, involves selling timber largely within the community. Table 8.1 summarizes the way in which those cases relate to our typology of community capabilities, market development and forms of market engagement.

The three sets of variables cannot be analysed in isolation and it is precisely their interactions that explain the diverse ways in which smallholders and communities interact in the markets, whether for timber or for NTFPs. Table 8.2 correlates the community (or smallholder) capability variables with market development variables, creating four forms of market engagement:

1 community forestry enterprises with high capabilities operating in well-developed markets;
2 smallholders and communities with low capabilities engaging with relatively well-developed market networks;
3 smallholders with little capacity and engaging in poorly developed markets; and
4 probably less common, communities with good capacities but marginal connections with markets.

Adopting this typology constitutes a useful way to assess specific cases.

Seven cases of community engagement with markets

Linking to markets through community enterprises

In the following two cases, community forest enterprises are linked to markets through formal contracts with timber companies, often with the mediation of an NGO, which provides technical assistance and helps build alliances between the community and the enterprises. Although these communities tend to obtain good financial returns from commercial logging, they also have little freedom to choose competitive markets. The two cases show that larger benefits can be obtained from the sale of sawn wood if the communities can surmount the technical and managerial challenges; otherwise they would remain providers of raw material for large-scale enterprises.

Layasiksa community in the North Atlantic Autonomous Region (RAAN), Nicaragua Indigenous and ethnic communities in Nicaragua's Caribbean Coast autonomous regions are gradually obtaining formal titles to their forests and other traditional lands. A 2002 law ensured formal recognition of rights to indigenous communities, but real interest in titling indigenous territories emerged only five years later, with the return to power of the Sandinista political party. Yet titling has been slow. Layasiksa sought and gained formal recognition in 1996 for part of the territory it claimed (35,000ha), but several

Table 8.1 *Factors shaping market participation in seven sites*

	RAAN (Nicaragua)	*Guarayos (Bolivia)*	*Northern Amazon (Bolivia)*	*Dolkha (Nepal)*	*Iturralde (Bolivia)*	*Porto de Moz (Brazil)*	*Nawalparasi (Nepal)*
Community capabilities							
Access to markets	Via logging road off main highway (periodically maintained dirt road)	Varies with proximity of community lands to main road	Depends on distance to main city and access to rivers or roads	Timber transport difficult, but some NTFPs are sold at roadhead	Varies; major settlements are close to main road	Timber markets distant; some buyers are in local urban centre	Via main national highway passing through village
Bargaining power	Logs are sold to company that pays slightly less than market price; sawn wood is negotiated and sold in capital	Little power; high dependence on financial capital from loggers, though power of negotiation has improved through collective forest management	Little power to affect price and compete with Brazil nuts gathered outside community lands	Low bargaining power; producers are far from market and farmer organizations are weak	Little capacity to negotiate with buyers; most operations are small-scale and informal	No power to negotiate price and sale conditions; relationships with local loggers are sporadic	Low bargaining power; heavy regulatory restrictions on sale of timber outside community
Knowledge of markets	Good knowledge of market options for logs and sawn wood; community hires sawmill services	Limited; no channels of market information exchange, little experience	Good knowledge; access to financial capital is provided through patron–client relationships	Poor knowledge; limited literacy, limited means of mass communication; distant from markets	Fragmented, limited to needs of buyers in specific timber transactions	Limited, with little opportunity to acquire information; market is poorly developed, timber demand is sporadic	Limited and skewed; reliance on small business operators for market information
Organizational capabilities	Good skills for managing forest resources, with help of outsider organizations	Still precarious but improving through creation of community forest enterprises	Varies with development of community organizations	Strong organization, good social capital and collective initiatives; naive in dealing with markets	Not apparent; logging operations are individual, mainly outside law	Weak; economic activities are carried out on family basis	Weak capacity to participate in market

Market development							
Process of price formation	Negotiation based on species available according to management plans, along with volumes and quality, which define final price	Highly influenced by local sawmills that finance community logging operations	Annual price negotiation, prior to harvesting season, involves industry, landholders and gatherers	Buyers set prices for NTFPs; prices very low compared with market price in Kathmandu	Price set largely by timber traders, with little influence from smallholders	Timber price set largely by local loggers in negotiation with regional industry	Price for internal sale set by CFUG leaders; exports restricted
Availability of information	Communities depend on support from outsiders to access market information and negotiate prices	Unequal access; information is transmitted through informal commercialization channels	Lack of information channels, little knowledge of market trends	Information on market and price is available largely through NTFP traders	Information is transmitted through informal and illegal networks	Access to market information is controlled by industry and traders	Information is transmitted through informal channels
Presence of buyers	Timber markets relatively well developed, with several regional companies and local loggers	Several buyers interested in buying timber from communities	Market controlled by several industries; large number of traders	NTFP market controlled by small number of traders linked to Indian markets	Several buyers, largely financed by capital from outside region	Market controlled largely by several buyers who finance informal logging operations	Domestic market controlled by government-owned corporation; small businesses operate locally
Market regulations	Export of logs and planks banned; no market regulations in internal markets	Exports of logs banned; no market regulations for internal markets	No market regulations for internal timber markets	Sales outside community members are heavily regulated	Exports of logs, banned; no regulations for internal markets	No market regulations for internal timber markets	Sales outside community are heavily regulated; high transaction costs
Forms of market engagement							
Product sold in market	31 timber species available for logging; species harvested depend on market demand	Timber species, as determined by market demand	Brazil nut, which is widely available	Mainly unprocessed NTFPs; essential oils are recent products	Valuable timber species largely extirpated in other forest regions	Several timber species with demand in regional market	Timber for household consumption, sold within community
Stability of transactions	Stable interactions in roundwood and sawn wood markets; logging only in three-month dry season	Annual sale from harvest, based on forest management plans	Seasonal market	Annual sale at time of collection	Sporadic engagement with market	Sporadic engagement with market	Annual, usually during winter

Table 8.2 *Types of engagement in forest markets*

Market development	*Community capabilities*	
	High	*Low*
High	Enterprises with stable integration to forest markets	Individuals or community groups with limited integration to markets
Low	Well-organized community groups not integrated to markets	Individuals with sporadic interactions in forest markets

years passed before the community could enforce its exclusive rights over that area and it is still negotiating certain borders with neighbours (Larson, 2008). The community of Layasiksa has created its own community enterprise, Kiwatingni.

The community developed two forest management plans and certified its forest operations with help from donors. Technical staff from the World Wildlife Fund (WWF), who had supported Kiwatingni, have established their own company, Masangni, which plays an important role in contracting, oversight and community training. Kiwatingni contracts with Masangni to obtain the services of a forester for the development and oversight of the annual operating plan. Layasiksa has relatively diversified forest operations, which range from log harvesting to sawn wood production. One of the management plans (covering 4950ha) is for a ten-year concession to the company Prada S.A., which owns a sawmill in the neighbouring municipality, and the other (covering 4664ha) is managed by Kiwatingni (Argüello, 2008). Both plans involve broadleaf forest.

The Prada concession is for the sale of standing timber, which is sold at US\$6/m^3; the community signed the contract in 2002 without any provision for renegotiating the price over the ten-year term and was hence forced to sell at this price, despite rising timber prices, until Prada agreed to pay US\$7/m^3 for wood harvested in 2008 (Larson et al, 2008). The community does not participate in any of the harvesting decisions, which have all been ceded to the company. Thus, community efforts are concentrated in the second area, where Kiwatingni makes all decisions regarding choice of species, harvesting techniques, percentages to sell as logs or as sawn wood and so on. Since it does not own heavy equipment or sawmills, Kiwatingni hires service providers for hauling, transportation and milling and supervises them closely (Larson and Mendoza-Lewis, 2009).

All the round wood is sold to Prada S.A., which pays a slightly lower price for logs than other buyers, but it pays cash on delivery, is not as strict about quality and purchases additional, less valuable species, since its main product is plywood. It also provides all the fuel required for Kiwatingni's operations (Argüello, 2008). The community is thus highly dependent on Prada S.A. The community also produces sawn wood by hiring milling services, and it has always worked with the same sawmill owner even though there are several local portable sawmills. The community hires local truckers to transport wood to Managua, the capital, as needed (Argüello, 2008).

Marketing has been carried out directly with clients rather than via traders. Masangni plays a central role in marketing. Masangni, WWF and other donors that support Layasiksa help negotiate prices, promote the use of lesser known species and lobby for the use of certified wood. Buyers are mostly in Managua, two days' travel from the community. There are no large buyers who would purchase all of the species, sizes and qualities sold, though a single buyer purchased 70 per cent of the sawn wood in 2007. There are no formal contracts, hence there are no legal obligations for buyer or seller, just a 'note' indicating volume, quality, species, dimensions, price and means of payment. When one buyer failed to pay by the agreed date, Kiwatingni had no legal recourse, but at the same time this arrangement gives the community more flexibility (Larson et al, 2008).

The sale of round wood generates a net financial loss, mainly because of the high cost of equipment rental from the service provider. This loss is absorbed by the gains obtained with sawn wood production, which has a net return equivalent to 21 per cent (Argüello, 2008). In 2007, Kiwatingni made a profit of about US$17,500, or 9 per cent, which is low compared with similar enterprises. This amount includes the costs of training and technical assistance as production costs, which were actually paid by Masangni. By the community's own accounts, which do not include these costs, they earned about US$30,000 in profit (Larson and Mendoza-Lewis, 2009).

This case shows that a lack of competition for service provision raises production costs; that inadequate access to capital leads to unfavourable dependent relationships with a large company, and that ongoing outside support, particularly with regard to marketing, is needed. At the same time, the community earned almost US$22,000 in wages. As a whole, this means that about US$0.43 on every dollar generated along the chain from planning to sale went back to the community (Argüello, 2008).

Community forest enterprises in Guarayos, Bolivia Guarayos is a province of Bolivia's Santa Cruz department, which is home to the Guarayo indigenous people and a rapidly changing forest frontier. In 1996, this indigenous group presented a territorial demand for 2.2 million ha to the government; areas totalling 1 million have already been titled (Albornoz et al, 2008). The remaining areas face intense pressure from medium- and large-scale landholders, who are also demanding clear tenure rights for their agricultural and livestock operations. A 1996 law initiated a process of land regularization (*saneamiento*) to clarify tenure rights for individual landholders and indigenous land claims. Furthermore, around 562,000ha of production forest was granted as concessions to 11 timber industries after the approval of the Forest Law in 1996 (Vallejos, 1998). The proximity of Guarayos to a paved road connecting the cities of Santa Cruz and Trinidad, urban centres in the Bolivian lowlands, has increased interest in the area's forest resources.

The indigenous communities wanted to develop timber management plans as a strategy for consolidating their hold on forest areas that were not occupied and were thus viewed as available to outsiders. From 2000 to 2004, six indigenous groups gained approval for management plans in forests around their communities and created community forestry enterprises, often with the help of forestry projects and NGOs. In total, 211,178ha of forest was placed under management plans, with individual plans ranging from 2,433ha to 60,000ha (Albornoz et al, 2008).

The community of Cururú formed the Cururú Indigenous Timber Association in 2001, based on a strategy for supporting communities developed by a USAID-funded forestry project, the Bolivia Sustainable Forest Management Project (BOLFOR) and other NGOs. The community comprises about 40 families, who set aside an area equivalent to 26,420ha for forest management (Albornoz et al, 2008). The forest management plan was approved in 2002, with a 30-year harvesting cycle. Since that time, the timber association has been undertaking annual forestry operations, which by 2009 constituted the most important economic activity of the community. In 2007, Cururú was able to certify its forestry operations under the Forest Stewardship Council (FSC), though this was possible only with the financial support of external donors and timber companies that buy round wood from the community and had already certified their custody chains.

In 2002, the association placed six species on the market and sold a total of 1030m^3. In 2003, it sold round wood to the companies SOBOLMA and Monteverde. Since 2006, it has increased its portfolio of buyers to include the main companies in the area – INPA Parket, La Chonta and CIMAL (Albornoz et al, 2008). CIMAL agreed to a five-year contract with the timber association, with prices and volumes to be negotiated on an annual basis. That has led to a significant increase in cash income from logging since 2007. Whereas in 2002 the community obtained a profit equal to US$14,900, it more than doubled in 2007, reaching about US$34,500 and generating a significant income stream for the families involved: US$1014 per family (BOLFOR II, 2007).

The influx of timber income has had significant implications for the community economy. People's livelihoods have shifted from subsistence agriculture and off-farm wage work to forest-based economic activities, allowing young men to remain in the area rather than migrating seasonally to work outside the community. However, tensions have increased over the distribution of benefits from the management plan. Income is sporadic, with cash payments often only available in the months following the annual harvests, sometimes after substantial delay. Also, at times, tensions have appeared between families of leaders, who have invested more and thus earn more from the project, and their neighbours (Albornoz et al, 2008).

Communities in non-timber forest products markets

The next two cases exemplify community engagement in the marketing of NTFPs. These markets feature a large number of suppliers and buyers, and patron–client relationships tend to dominate market transactions. Low entry costs encourage many individuals to supply small amounts of product, resulting in huge supplies and sometimes attracting a large number of buyers organized in complex networks. These networks are ultimately controlled by a few buyers and industry.

Brazil nut cooperatives in Bolivian northern Amazon The Bolivian northern Amazon is a remote forest frontier now connected to the rest of the country by a new road. Historically, NTFPs have been the basis of the region's economy. Initially, in the late 19th century, occupation of the region was driven by the rubber boom but later shifted to other NTFPs. Brazil nuts (*Bertholletia excelsa*) have been a principal NTFP extracted from Bolivia's northern forests since the mid-20th century and more recently have become the foundation of the regional economy (Stoian, 2000). In fact, since 2003 Brazil nuts have been one of Bolivia's more important forest exports. Brazil nuts are collected in the forest during the rainy season, from December to April. Nuts left in the forest deteriorate rapidly.

Rural forest concessionaires, private landowners, *barraqueros* (the holders of *barracas,* the units of forest exploitation in public forests) and smallholders participate in this economy. Smallholders have recently gained communal property rights over large expanses of tropical forest and may account for a third of total production (Bojanic, 2001). Their rights are based on customary claims to territory traditionally used for nut gathering on lands that were claimed until recently by *barraqueros* (Assies, 2008). Independent communities now hold lands adjacent to those of *barraqueros*, who continue to control some land and trees and organize production in enclaves with wage labourers.

Brazil nut collection constitutes one of the most important sources of income for communities today, supporting families throughout the year (Stoian, 2005). Local markets are governed by a complex network of informal but highly developed institutions. Entry is relatively easy, with household-scale production and low capital inputs. The rise of Bolivia's processing industry has been crucial for the growth of the Brazil nut sector. Some companies own *barracas*, but most finance *barraqueros* and traders to secure access to raw materials for their factories. *Barraqueros* use such financial resources to hire seasonal workers to collect the Brazil nuts on their lands. In turn, communities sell their production to traders, from whom they receive small cash advances to begin gathering nuts.

Producer cooperatives have increased the influence of smallholders in the commodity chain, as well as their participation further along the chain. For example, the COINACAPA forest cooperative formed in 1998, with 41 members (see Chapter 6). In 2008, it had 465 members from 40 northern Amazon communities. Members are not required to sell all their production to the cooperative and many sell a portion to local traders to maintain a relationship with them as well as to spread the risk (Albornoz and Toro, 2008). The harvest is placed in crates that can hold 20kg of nuts. On average, a smallholder collects about 200 Brazil nut crates per year at a price of $8.50

each, making a total of US$1700 per year. Brazil nut prices are negotiated every year among representatives of the processing industry, *barraqueros* and seasonal workers. International prices skyrocketed in the middle of the first decade of this century but then fell rapidly as a consequence of the international economic downturn.

COINACAPA contracts a processing plant for shelling (rather than exporting them with shells), then exports to Fair Trade and organic brokers. Currently, the cooperative produces 112 tons annually. The prices obtained in the organic market are a bit higher, about US$0.10 to 0.15 per pound. Producers are paid the local market price at the time of harvest, then COINACAPA distributes profits after processing and sale. This has usually meant an additional payment of about 50 per cent of the original market price and the second payment arrives several months later – just as cash income generated during the harvest is running out.

Suspa community forest user group in Dolkha, Nepal The Suspa community is in the Dolkha district, in the mid hills close to the Himalayas about 100km northeast of Kathmandu, the capital city. Dolkha's altitude ranges between 2500m and 4000m above sea level and the terrain is very difficult. Only one road traverses the district and it is more than two hours away. The community depends predominantly on subsistence farming combining crops, livestock and forest management (Banjade and Paudel, 2008). Suspa's 303 households have been managing 645ha of forest since 1995. All households in the settlement are members of the CFUG. One member from each household is represented in the general assembly. A 15-member executive committee carries out everyday management decisions and all its decisions must be endorsed by the general assembly.

The forests here are rich in both timber and NTFPs. The community is dependent on the forests for construction material, grazing, fodder, fuel wood and many agricultural implements and wooden household utensils. The timber sold in 2006–2007 to members of the community was worth US$2228 (Paudel et al, 2008a). The CFUG has a decent stock of harvestable timber and a good market, especially in the district headquarters (two hours' walking distance away). Nevertheless, forestry regulations do not account for the complexities of high-altitude forests. For example, collection fees and high transportation costs make it expensive to take unprocessed timber to the market, but there is a prohibition on establishing sawmills inside the forests. Altogether, the forest contribution to members' household income is relatively small, 13 per cent of the total (Paudel and Banjade, 2008a).

The collection and sale of NTFPs is managed by the authorities in a relatively relaxed manner. CFUG members can individually collect and sell NTFPs to traders. Most people involved in timber harvesting operations as paid labour also collect NTFPs, including lokta (*Daphne bholua*) and argeli (*Edgeworthia gardeneri*) for the paper industry. The paper industry sets the price for these raw materials. In some cases steamed and shelled argeli is sold to a trader in Kathmandu who finally exports it to Japan. Other NTFPs are mushrooms, which are sold in the district headquarters, and wild vegetables. The value chain for these products is relatively complex: dozens of traders stand between the collectors and the large merchants in India. A study on the status of NTFP businesses across Nepal indicates that the primary collectors get less than 12 per cent of the consumer price (Subedi, 2006).

The CFUG has planted cinnamon species in the community forest (Paudel et al, 2008a). Again, the outside traders set the price and often complain about the quality of the products to justify the low price.

Suspa is also involved in an essential-oil extraction plant for processing wintergreen (*Gaultheria fragrantissima*), which is locally known as machhino. The CFUG collaborates with the Deudhunga Multipurpose Cooperative Ltd in running the processing plant (Gurung, 2006) and many of its members are employed there. Collection of machhino is labour intensive. It has been reported that members supplying raw material to the processing plant make about 17 per cent more than the average income. Total household income averages US$250, of which US$43 is generated by this community enterprise (Acharya, 2005).

Informal interactions on market fringes

Many timber markets in the tropics are sporadic and informal, or even illegal. Communities in this type of situation do not maintain a stable relationship with the market and they do not always follow the regulations governing their forest resources. A major portion of the timber sold by these communities does not originate in authorized areas. In the following two cases, market traders who have access to both financial capital and market information often have the advantage.

Colonists in Iturralde in northern La Paz, Bolivia Iturralde is a province of Bolivia's La Paz department. Intense logging occurs in this province, where a large percentage of the timber is harvested informally. It originates in small-scale plots owned by colonists who migrated to the region a few years ago. The colonists rarely obtain the required forest management plans, which in theory would grant them legal rights to use the timber on the land. But even if they were willing to pay for developing such plans, many would be unable to do so, since they do not hold formal ownership rights to the land they occupy (Pacheco et al, 2008b). Colonists log the trees not only from their lands but also from other areas, primarily in the neighbouring Madidi National Park, a protected area created in the mid-1990s.

The expansion of colonist settlements around the urban centre of Ixiamas, the capital of the province, has inevitably led to the expansion of logging activities in the region, even though most received (or informally occupy) 50-ha plots intended for agricultural activities. Timber companies and small local associations, both with access to forest concessions in the area, create the demand for timber. This has also led to a growing number of sawmills, some operated by large timber companies. Approximately a fifth of the logging permits in the region have been granted in colonization settlements (Ibarguen, 2008). The timber supplies several hundred small-scale sawmills and carpentry establishments, along with a few large-scale wood processing plants (Solares, 2008). Ibarguen (2008) suggests that logging – both selling trees and wage labour – constitutes an important source of smallholders' income.

About 900 colonist families totalling 4500 people occupy a total area of 75,000ha under individual and collective ownership. Land-use decisions are made mostly on a family basis, including those regarding forest resource use. Most of the harvested wood originates from areas without legal permits and is sold to local traders who send it to the city. These traders find ways to deliver the illegal timber to the final buyers.

Illegal timber trade takes place along the whole value chain, from harvesting to transportation and processing. Many timber companies with legal operations in forest concessions buy illegal timber, as do local traders who may be financed by medium-scale processing companies or timber export agencies based in the capital. Smallholders constitute only a portion of this intricate network. In some cases, local loggers finance the formulation of forest management plans in community lands but use them only to obtain forest permits and then sell them – a more profitable activity than the forestry operations themselves. According to Ibarguen (2008), seven traders have the capacity to advance money to finance logging operations, the equipment for harvesting and hauling and the information about buyers, as well as the best ways to avoid the control points established by the forestry agency. These traders often mobilize resources that promote illegal logging inside the Madidi National Park and other public forests.

The available figures suggest that illegal logging on public lands is more profitable than legal logging in smallholder plots. A smallholder's profit depends on the species and volume and also on the ability to negotiate. Ibarguen (2008) finds that the average income from forest management in a 35ha plot is about US\$8/m^3, for a total of US\$2800 on a 350m^3 harvest. Profits from illegal logging of mahogany, however, come to US\$165/m^3. Smallholders can make more money by encroaching on the national protected area.

Traditional communities in Porto de Moz, Brazil The timber market in Porto de Moz is characterized by high transaction costs for undertaking forest management through formal plans and the influence of local traders who have the capital and information to operate, but it is isolated from the main market, located in the city of Belem. Porto de Moz is on the Lower Xingu River in the northern part of the state of Para. The occupation of community lands by large-scale timber companies in the 1980s and 1990s led to intense land conflicts with local communities (Moreira and Hébette, 2003). A broad-based movement allying resident communities and environmental NGOs was successful in drawing attention to the region and in 2004 the extractive reserve (RESEX) Verde para Sempre ('Green Forever') was created.

The RESEX forced out the timber companies working in the reserve and granted land rights, with restrictions, to smallholders living there. The establishment of the RESEX resulted in the restructuring of local timber markets. Local loggers became more politically powerful and informal timber markets expanded. The shadow networks that existed previously continued to operate but with different sources of capital. The new traders moved logging pressures to the eastern side of the reserve, not only to the community lands there but also to national forests, such as the Caxiuanã National Forest (Nunes et al, 2008).

Logging in community lands on the eastern side of the RESEX is intensive. Smallholders are approached by local sawyers interested in a few valuable trees; these sawyers in turn are financed by traders, who also pay logging truck owners (*bufeteiros*). In Porto de Moz there are three local traders of round wood and another three interested in sawn wood, though it is also possible to find buyers from surrounding municipalities. These buyers lack formal access to forest areas and often operate in the shadows. Furthermore, there are three large-scale sawmills in the area, which sometimes use local buyers but also make use of their own networks for timber supplies.

Information on the amount of timber sold in Porto de Moz is unavailable, but Nunes et al (2008) provides figures on costs and benefits from informal logging. Smallholders often sell standing trees only, for US$13 each, because logging costs (including oil, labour and equipment) are unaffordable. A local chainsaw operator makes around US$70 for harvesting the tree and selling it to the local intermediary, but the sale price increases to more than US$200 per tree when it is converted and sold as planks. Smallholders have little scope for price negotiation; sawyers can negotiate prices with traders but are dependent on cash advances from these same traders for their own forestry operations, a factor that limits their bargaining power.

Timber transactions without open markets

The last case is another community forest user group in Nepal, which sells timber largely within the community – not only because the community forest is intended to meet the needs of the members, but also because regulations make it difficult for forest users to sell timber in the open market. It is not that open markets do not exist, but rather that communities cannot reach them.

Forest community of Sundari in Nawalparasi, Nepal The Sundari CFUG is in southern Nepal, in the lowlands, and is linked with the national highway. The area has an elevation of 650–700m above sea level and lies in the tropical zone. The forests are important in terms of both commercial values and biodiversity. Some high-value medicinal species grow in the region, as do lemon grass, French basil, asparagus and more than 200 other species valued for their fruits and seeds, root and rhizomes, bark, leaves and flowers (Paudel and Banjade, 2008a).

The community comprises 1216 households and manages 384ha of forest. The community relies mainly on subsistence farming and remittances, although the poorest people depend mostly on agricultural wage labour (Paudel et al, 2008a). The Sundari CFUG does not sell timber in the market even though some 20 sawmills operate in the area. These mills primarily buy timber from the Timber Corporation of Nepal, a parastatal company. A minor portion is supplied by CFUGs, however, usually through auction. Most of the mills provide sawing services and make furniture for the local CFUGs (Paudel and Banjade, 2008a).

The Sundari CFUG has a dense subtropical forest with valuable sal (*Sorea robusta*). The price of sawn timber in the open market is US$536/m^3, but the CFUG sells at only about 25 per cent of the market price to its members. The CFUG manages the forest based on an operational plan, a contractual document signed by the district forest officer. The group harvests timber once a year, often during winter, and distributes the timber and fuel wood among the members for a nominal charge, which it determines. Individual members of the CFUG are not allowed to harvest timber on their own.

One of the main regulations on community forest use is that all members have access to fuel wood, fodder, leaves and medicinal plants, but only for household use, not for sale. The group has agreed on a rule that allows only the CFUG, not individuals, to collect and manage these products. The group also sells some NTFPs, but in the fiscal year 2006–2007, it sold less than US$375 in NTFPs. In the surrounding area, about ten NTFPs from the community forests are traded in local markets, which are dominated by Indian traders. The products are mainly collected at the roadheads or in the local weekly market (Paudel and Banjade, 2008a).

The CFUG harvests timber according to the volume limits set by the operational plan. The overall harvest amount is negotiated every year and has been more or less the same over the years. In 2008 the Sundari CFUG harvested 170m^3 of round wood, though internal demand exceeded 510m^3 (Paudel and Banjade, 2008a). The harvested timber is sold to members based on their need for it. Taking equity issues into account, the CFUG has set prices based on members' wealth category – poor, medium, rich – at US$100, US$120 and US$145 per cubic metre.

Cumbersome bureaucratic procedures apply if the group decides to sell any timber outside the community. For example, it must first publicly offer the timber for sale to all neighbouring communities in the district. The group can access the open market only when the timber stock remains unsold at the neighbourhood and district levels (Bampton and Cammaert, 2007). CFUGs are supposed to receive the total amount from any timber sale but are liable for a 15 per cent royalty to the government on sales outside the forest user groups.

The constraining regulatory framework has two important consequences. First, it discourages sale of timber in the open market, resulting in low timber rents for the group as a whole (Paudel and Banjade, 2008a; Bampton and Cammaert, 2007). The low price then leads to increased demand for timber, particularly among the wealthy members in the community, and some timber may be illegally leaked into the local sawmills. Second, it worsens the existing inequity in distribution of conservation benefits among the members, through a phenomenon called hidden subsidy (Iversen et al, 2006). The better-off

Table 8.3 *Categorization of case studies' engagement in forest markets*

Market development	*Community capabilities*	
	High	*Low*
High	1. Communities with stable forest market integration (timber): Layasiksa in RAAN, Cururú in Guarayos	2. Groups with limited market integration (NTFPs): Bolivia northern Amazon, Dolkha
Low	3. Individuals with sporadic forest market interactions: Iturralde, Porto de Moz	4. Well-organized community groups with little market integration: Nawalparasi

members within the group consume more than 80 per cent of the cheap timber. The poor hardly buy any timber, even at this low price. Thus the low price set by the CFUG largely benefits the wealthy at the expense of poor members.

Discussion

The seven cases can now be assessed according to the typology introduced above (see Table 8.3).

1 Communities with stable forest market integration The two cases show progress in community attributes for dealing with forest markets, probably the result of the intervention of international NGOs. Both communities have improved access to markets by enhancing knowledge about buyers, along with market conditions such as prices, volumes and qualities. However, it is not quite clear yet to what extent the communities are the direct actors in negotiating deals with companies, or if the forestry projects still act on their behalf, or if they would have the capacity to compete in timber markets if subsidies were removed. In both cases, communities have improved their incomes by becoming more active in the market, mainly by establishing alliances with timber companies, which are reliable buyers. Nevertheless, the community forestry enterprises have little scope for negotiating the prices for timber and other conditions established in the contracts, such as payment conditions, because of their dependence on the companies.

2 Groups with limited market integration (NTFPs) The next two cases are smallholders and communities with strong ties to NTFP markets. These products constitute important sources of household income. Although smallholders have fairly good knowledge of prices and qualities, they often depend on patron–client relationships with buyers. For example, smallholders in the Bolivian northern Amazon need cash advances in order to cover their everyday costs during Brazil nut collection. Cash advances are not unusual in Dolkha, either, where Indian traders advance sums to NTFP collectors, thereby binding the individual collectors to them. Commonly, however, community members are paid in cash. In the two situations described, smallholders are not able to negotiate the price of their products. The two cases demonstrate that organizing for collection, processing and marketing makes a difference. Cooperatives can build alternative channels of commercialization and even reach international markets – mainly fair trade and organic. Such enterprises may still depend on financial resources from traders or on credit provided by microfinance agencies, acquired at high interest rates.

3 Individuals with sporadic forest market interactions The two cases illustrate how smallholders can have relatively good access to markets but lack the bargaining power to enhance their form of engagement because of weak organizational capacities, individualized access to forest resources and small-sized plots. Furthermore, they often lack formal titles to their lands and encounter high transactions costs to obtain forest management plans as well as meet unrealistic forestry norms. These are all factors that encourage smallholders to engage in informal and illegal relationships with timber buyers: the former cannot comply with the regulations and the latter tend to benefit from that situation. Although the smallholders avoid some transaction costs, this type of market interaction works against them in many ways, mainly by undervaluing the standing trees or round wood that they offer in the market and inhibiting them from negotiating fairer prices.

4 Community groups with little market integration The last case is a community that has good capabilities but maintains little relationship with open timber markets. One reason is that the community's needs for timber exceed its supply. The second reason is that cumbersome forestry regulations hinder the community's efforts to sell its timber outside the community. In this case, the benefits that communities can obtain from these markets are relatively limited, since the timber tends to be undervalued in order to fulfil social needs. The low price set by the CFUG for its members reduces timber rents, which limits the group's capacity to invest in community infrastructure and other social services. And since better-off members buy more cheap timber, they enjoy an indirect subsidy from the rest of the members.

Implications

It is clear that business-oriented, community forestry enterprises have helped some communities engage in more formal and stable relations with timber markets, thereby providing a regular source of income from commercial logging. By engaging in the marketplace, community enterprises slowly improve their negotiation skills for making deals with buyers, hiring service providers and compromising with government officials. In many cases, the approval of a forest management plan, which is the first step in formal forest management, leads the community enterprise to maintain permanent relationships with the markets. But to obtain benefits from markets, many communities must first meet legal requirements, obtain financial resources and improve their accounting and marketing skills. Communities often require the support of external projects and NGOs.

Communities tend to obtain greater benefits to the extent they are able to sell their produce in more competitive markets and under more transparent conditions. Selling roundwood may be a trap for communities in monopsonic markets, since they become price-takers and are at the mercy of traders and timber companies. Selling sawn wood would reduce transportation costs and simplify access to more distant markets in which the sellers could make more attractive deals. Not all community forestry enterprises can set up their own sawmills, however, in part because of regulatory and bureaucratic constraints. For example, in Nepal, no forest-based enterprises can be established within a 1km periphery of the forest and thus the communities cannot add value to their forest products. Some smallholders operating in informal markets use chainsaws to produce planks, which often receive a higher price in the marketplace.

How profitable are community forestry enterprises? Net profits are about US$30,000 in Layasiksa and US$34,500 in Cururú. The profits per family, however, amount to only US$177 and US$1014, respectively (Albornoz et al, 2008; Argüello, 2008). In contrast, a smallholder in northern La Paz can net about US$2800 by extracting the valuable trees from a 35-ha lot (Ibarguen, 2008), and in Porto de Moz, sawyers can reach a monthly income of about US$1000 for sawing five trees into planks (Nunes et al, 2008). In the Bolivian northern Amazon, incomes from Brazil nut gatherers depend on the availability of this resource and may range between US$320 to US$2000 (Albornoz and Toro, 2008). In Dolkha, Nepal, the CFUG obtained about $3200 from timber sales in 2007 (an equivalent of US$11 per household) and families collecting NTFPs from the community forest made from US$8 to US$43 during the same year (Banjade and Paudel, 2008).

Even though members of community enterprises obtain less annual income from legal, commercial logging, they can count on a regular income over many years. Smallholders, in contrast, often receive higher amounts for their valuable trees, but in a lump sum, and the valuable trees are disappearing. Smallholders gathering Brazil nuts obtain the largest benefit but are highly dependent on international prices. Finally, it is noteworthy that communities in Nepal see lower profits than the cases in Latin America, for several reasons: Nepalese population densities are higher, the community forests are smaller, the forest resources generally have lower value and forestry regulations are tighter.

No clear pattern emerges regarding the implications of the different forms of community engagement with markets for forest condition; too many other factors are involved (see Chapter 9). However, timber extraction tends to alter the forest conditions for NTFPs, whose collection contributes toward forest conservation. Formalization of forestry operations and the creation of community forest enterprises tend to have a positive impact if they introduce sound forest management practices. But forest enterprises may also engage in large-scale extensive forestry operations, which in the long run tend to have a detrimental effect on forest regeneration. Informal and illegal markets can erode forest resources to the extent that more intensive logging is practiced, but their effects tend to be more limited, since the operations are small and often just one part of forest resource use. This last issue, however, requires more in-depth research.

Lessons and policy prescriptions

Although smallholders and communities have gained benefits from tenure reforms, even secure tenure does not guarantee that local people will benefit from the commercial use of their forest resources, since these benefits depend on both community capabilities and the conditions of market development. Tenure reforms do not affect the structural conditions under which forest markets operate or the interactions between smallholder and communities with traders, intermediaries, sawmill owners and industry. Market conditions and relationships play a major role in explaining how much benefit smallholders and communities are able to obtain from the commercial use of their timber and NTFPs.

The cases presented in this chapter suggest that distortions in forest markets often work against smallholders and communities. Timber markets tend to be dominated by a few companies that exert great influence on pricing, and even at the local level relatively few buyers may determine prices. In many cases, regulatory constraints tend to stimulate informal logging, which lowers timber prices. Transactions in the NTFP markets are often shaped by patron–client relationships with asymmetric information, which tends to disadvantage collectors. The market distortions, in several cases, inhibit both community capabilities and further market development.

The challenge for enhancing smallholder and community benefits from commercial resource use, then, is twofold: first, to enhance community capacities in the markets and second, to modify the conditions under which such markets work in practice. The former issue is part of the agenda of donor and forestry projects; the latter issue is often neglected by public policy.

Community capacities can be enhanced by promoting enabling environments for developing entrepreneurial initiatives, rather than by implementing interventions based on outside models. Demand-driven approaches for improving access to technical services and learning exchange networks to disseminate knowledge and experiences are examples. Furthermore, communities need tools to better understand market functions and trends and the present and future value of their resources, so that they can make more informed decisions.

Public policy action is required to overcome the main structural market distortions highlighted in this chapter – especially the legal and regulatory barriers, patron–client relationships and asymmetric information. Forest regulations that constrain communities from using their forest resources have to be relaxed and adjusted to forest user needs. State forestry agencies and other forestry programmes supported by donors should build stronger alliances with communities so that they can participate more actively in forest markets. Market asymmetries can be addressed by making financial and other technical services available to communities through more flexible intervention schemes. Although state intervention through public companies (such as the Timber Corporation of Nepal) can be a good option, these companies must be accountable and their operations transparent and efficient. State agencies could also intervene directly by mediating in market transactions to develop emerging markets, particularly for certain NTFPs.

A final, perhaps obvious, lesson emerging from our analysis is that policy interventions cannot be the same for the different forms of engagement. In some cases, public policy should focus on supporting community capabilities to interact in markets; in others, greater attention should be placed on affecting market conditions. In many cases, the two types of issues should be given equal priority.

Note

1. This chapter explores the interaction of both smallholder and communities in markets, though in some cases it is difficult to draw a precise line between these actors. In general, the term community is used in a more generic sense to depict situations that embrace both community groups and smallholders, but also to refer to situations in which group decision-making predominates. Instead, the term smallholder deals primarily with individual farmers using their own private resources.

Part V
Outcomes and Conclusions

9
Outcomes of Reform for Livelihoods, Forest Condition and Equity

Ganga Ram Dahal, Anne M. Larson and Pablo Pacheco[1]

Are tenure reforms improving local people's livelihoods and conserving the forest? To what extent is it possible for tenure reform to achieve the two goals simultaneously? What are the implications of these reforms for equity? This chapter provides insights into these questions by assessing the outcomes of the reforms in our case studies and discussing why and under what conditions they have resulted in improvements or deterioration.

Livelihoods, forest condition and equity are also affected by other changes, broadly associated with increasing urbanization, agricultural development, industrialization and technological transformation. Tenure reform is only one of several processes shaping outcomes. In this chapter, however, we try to isolate the effects of the reforms, based on the assumption that the nature of tenure rights – e.g. tenure security and the specific locally relevant components of the 'bundle of rights' – shapes the decisions people make concerning forest resource use. The reforms, therefore, are likely to have significant implications for the livelihoods of people who depend on forest resources and for the ways in which forest resources are used.

There is substantial literature, only briefly alluded to here, on the associations between security of tenure and improvements in livelihoods and incomes. But much of this literature (e.g. Deininger and Binswanger, 2001) is based on situations regarding private, individual titles – not the norm in forest tenure reform. More relevant for forests is the common property literature, which has consistently demonstrated that livelihood benefits are more likely to result from secure common property rights (Pagdee et al, 2006). Still, there remains uncertainty about the actual benefits, for both communities and forests, which derive from greater tenure security. Indeed, this uncertainty has been one of the stimuli for our research.

In theory, tenure reforms with conservation goals will achieve better impacts for forest conservation but weaker livelihood outcomes and, conversely, reforms aimed at enhancing livelihoods should have better livelihood outcomes but perhaps weaker or even detrimental effects on forests. But actual outcomes are mediated by several other factors. The first is the extent to which the reform has been effectively implemented and the political and economic context in which it takes place. The second, which depends in part on implementation, is the extent to which tenure reform increases rights in practice.

New rights that are not secure result in little change in rights or may increase insecurity under certain circumstances, such as with the imposition of statutory over customary rights (Mwangi and Dohrn, 2008; see Chapter 4) or with state interference in areas where communities had previously managed their own lives (Edmunds and Wollenberg, 2003). Hence the nature, goals and implementation of the reform affect the extent to which rights have increased and been secured in practice. Outcomes for livelihoods and forests should then be assessed in this light.

We seek to assess the results, on the ground, of tenure reform in more than 30 sites around the world with due humility. We recognize that even when reforms increase rights and tenure security, numerous other variables mediate outcomes. In fact, the study was designed to identify and understand some of those mediating variables and this book attempts to examine them more closely.

Throughout this research, we have first tried to ensure that our results are appropriate to local contexts. Then, in an iterative manner, we have developed methods that assess changes in livelihoods, forest condition and equity across the sites. This has involved ongoing iterations and communication among the field teams, throughout the two-year process, to maximize comparability. The field teams themselves are composed of people with long experience in their regions. The methods have included formal surveys, participatory rural appraisal tools, use of secondary data, interviews, maps and in some cases remote sensing data. The diversity of field situations required the teams to be creative and the results must be acknowledged to be qualitative. Most importantly, however, they are firmly grounded in local realities and the result of regular communications and adaptation among the authors and other researchers.

In general, we found that in many cases where communities have won substantial new rights, either forest conditions or livelihoods (and sometimes both) have improved. Most of the reforms have resulted in livelihood improvements, at least to some degree. Forest conditions were much more likely to improve when the reform included conservation goals. Though the data available do not make it possible to quantify the relative magnitude of forest and livelihood changes, and despite some cases with positive results for both, trade-offs between the two are evident. For example, in several cases, there were severe livelihood restrictions early in the reforms to favour improvements in forest condition, though these were partially lifted over time.

Where regulations and conservation restrictions were more severe, livelihood contributions have been more limited. Livelihood improvements are also limited because the forest resources being transferred to communities often do not include high-value resources, like the *terai* of Nepal or the high-quality commercial forests of Cameroon. There are also cases where forest conditions declined but livelihoods improved. Most importantly, the complexity of variables demonstrates the importance of understanding each case in context.

With regard to equity, the research found that at times, some of the poorest social groups suffered from new restrictions on resource access, but also, in a few cases, the participation of women and disadvantaged groups had increased. At a few sites, the poor were getting special consideration in forest products distribution, though these practices remain incipient.

Spheres of analysis

The rest of this introduction briefly presents each sphere of analysis (livelihoods, forest condition and equity) and the methods used. This is followed by discussions of findings across the sites and a short conclusion.

Livelihoods and income

Forests contribute to rural households through subsistence, with 'food, energy, medicine, fodder, housing, furniture, baskets, mats, dyes, agricultural implements', as well as for erosion control, inputs for soil fertility, pollination, weed and pest control and maintenance of water quality (Kaimowitz, 2003b, p46). Forests also contribute through small-scale sales of timber and non-timber forest products (NTFPs), community-based enterprises for forest products, wage labour and payments for environmental services (Scherr et al, 2002; Angelsen and Wunder, 2003; Sunderlin et al, 2005; Kozak, 2007; Molnar et al, 2007). Sunderlin et al (2005) identify four ways in which forest-based poverty alleviation can occur: through forest conversion, access to forest resources (by protecting current benefits or redistributing access and benefits to rural people), payments for environment services and increased value of forest products. Some authors argue, however, that forest conversion is more likely to overcome poverty.

In this research, simple qualitative parameters were used to assess livelihood outcomes. These include the assessment of the increased availability, as a result of the reform, of forest resources for the following basic elements of subsistence: place (*asentamiento*), shelter (house or home), food (subsistence agricultural production, hunting, gathering, fishing) and water (for all basic purposes).

Income was taken as one of the elements of livelihoods and measured in terms of relative shifts in income from forests over time, perceived shifts in total and relative forest income and specific, new forest-related income at the time of the study (with some attempt to determine whether any income losses were also associated with the tenure change). Incomes, as well as other livelihood changes, were generally measured at the community, not the household, level, since the community was the basic unit of analysis (see Chapter 1).

In most cases, researchers used focus group discussions, key informant interviews and the review of available documents to assess livelihoods and income changes. In all the cases, researchers had extensive previous experience in the sites studied, which improved both the quality of information gathered and their ability to analyse it in context and over time.

Forest condition

Forest condition may improve or decline with reforms. There are numerous examples of situations in which local people conserve or destroy local forests. For example, common property researchers have repeatedly demonstrated how local people organize and create effective local institutions to manage and conserve common forests, whereas Tacconi (2007a) and others have pointed out that sometimes forests may better contribute to livelihoods through clearing and conversion. As Agrawal and Chhatre (2006, p164) argue in the conclusion to their statistical analysis of 95 cases in India, examining causal factors related to forest condition, 'It may be impossible to identify a set of necessary and sufficient conditions' for effective local resource governance.

Forest condition was measured using three main variables. These include changes in forest cover over time, such as from digital maps at two points in time or, when these were not available, through a variety of interviews; changes in forest quality, through indicators of the increase or decrease in forest resource availability (e.g. specific plants or animals), and frequency of forest fires, also from official data or interviews.

Equity

If communities are benefiting in new ways from forest resources, how are these benefits being distributed? Communities are internally differentiated (Agrawal and Gibson, 1999) and access to resources may not be equitable (Ribot and Peluso, 2003). Research on equity in devolution or decentralization policies has often demonstrated problems with elite capture and the failure to include women, minorities and the very poor. Edmunds et al (2003) found that new management institutions created through state policies often opened up new income opportunities for elites and closed them for the poor. In Indonesia, ethnic conflict increased under forest decentralization policies as certain groups with connections rushed to take advantage of new opportunities at others' expense (Barr et al, 2001). Sarin et al (2003) argue that disadvantaged groups need 'explicit recognition of unequal gender and power relations, and firm provisions to ensure that livelihood interests and the rights of the poorest are given priority and protection'.

Equity was examined along two principal dimensions. The first involved determining who was considered 'in' the community of beneficiaries and who was considered 'out'. In some cases, the recognition of one group's rights led to restrictions in another's access. The second dimension refers to differentiation inside the beneficiary community. Gender equity was captured and assessed in different ways, including membership in forestry organizations, participation in leadership institutions, rules of inheritance and so on. In many cases, equity was also assessed along other dimensions, such as distribution of jobs and income and participation of ethnic minorities, the poor, migrants and youth.

Case studies

Our cases have been grouped such that they refer to a type of reform or a region, and most often both. This means that information from several community studies is being aggregated in some cases, and each case is, of course, embedded in its national institutional, economic and policy contexts, only some aspects of which will be discussed in relation to livelihood and forest outcomes.

Table 9.1 classifies the cases along two dimensions: the effectiveness of the implementation of the reforms and the relative weight of conservation and livelihood goals. By effective implementation, we refer not only to the establishment of a policy but also to associated laws and regulations and demonstrated progress through identifiable steps towards meeting the reform's goals. Clearly, a poorly implemented reform focused on forest dwellers would be expected to lead to fewer changes in rights, and possibly even increase tenure insecurity. Hence this should be taken into account in the assessment of outcomes.

Most of the reforms have been or are being fairly effectively implemented, though this does not mean they are problem free. For example, in the North Atlantic Autonomous Region (RAAN) of Nicaragua, demarcation and titling of indigenous lands made almost no progress for several years, but a new government administration is taking it much more seriously. The reform process at the Trans-Amazon site in Brazil has also been mixed, with implementation in older settlements more effective but with less progress in newer settlements. Reforms in Burkina Faso have involved ongoing tensions between statutory and customary rights, and the implementation of management plans at the scale of the concession is highly varied. Some of the most effectively implemented reforms are older projects, such as the community forestry sites in Nepal, the ancestral domain site in the Philippines, the Petén community concessions in Guatemala and the tree grower cooperatives in India.

Some sites have had serious problems in implementation or have simply never gotten off the ground. The community forest management model cases in the Philippines have suffered numerous stops and starts, with licences periodically cancelled; the programme itself has been suspended four times. Recognition of communal lands in the highlands of Guatemala has seen little progress, though two of the four sites studied are important exceptions. In Cameroon, community forests have been extremely difficult to establish without substantial outside support; in practice, private individuals have put up the funding and then also usurped the benefits intended for the community. The extractive reserve (RESEX) in Porto de Moz, Brazil, was established but then the development of the required management norms stagnated in the central

Table 9.1 *Case studies, by implementation and goal of tenure reform*

	Tenure reform goal		
Implementation	*Conservation, environment*	*Conservation, livelihoods*	*Livelihoods, rights*
More effective	CBFM, Nepal	KEF, Philippines Petén, Guatemala TGCS, India Concessions, Burkina Faso	Pando, Bolivia RAAN, Nicaragua Trans-Amazon, Brazil
Less effective or ineffective	Highlands, Guatemala	CBFM, Philippines Porto de Moz, Brazil Tree planting, Ghana	Guarayos, Bolivia Income sharing, Ghana CF, Cameroon

Source: Elaborated from country and site reports and discussions with researchers

bureaucracy. In Ghana, both types of reforms have implementation problems. The Modified Taungya System, presented here simply as 'tree planting', has no significant legal framework to give institutional validity to the reform; the other reform refers to community benefit sharing from logging contracts, but who receives the benefits or how these funds should be spent has not been clarified. In Guarayos, Bolivia, the slow progress of the reform, including delays of more than a decade, has increased competing demands on the most populated lands in the Guarayos indigenous territorial claim.

The second dimension in Table 9.1 is the primary goal of the reform, whether it aims to solve conservation and environmental problems or improve livelihoods or resolve rights demands, or a combination of these. Conservation clearly overshadows livelihoods or rights in two cases, at least in the goals and initial implementation of the reform: those include community forestry in Nepal and the communal forests of the Guatemalan Highlands. Community forestry in Nepal was primarily aimed at forest protection and initially included important restrictions on resource use. In the Guatemalan Highlands, many communities as well as governmental and non-governmental environmental agencies have prioritized resource protection over all other forest uses, particularly for the protection of water supplies.

Livelihood goals and rights are the primary driving forces in the two Bolivia cases, the Trans-Amazon of Brazil, Nicaragua's RAAN and the benefit-sharing arrangement in Ghana. This does not mean conservation concerns are irrelevant – in every case, certain regulations apply. Only in Guarayos and Pando, Bolivia, and in Nicaragua are reforms virtually unencumbered by conservation-based regulations, aside from requirements to have an approved management plan for logging. The remaining cases represent a combination of both priorities.

Several sites are located in official conservation or protected areas. The community forest concessions in the Petén are in the Maya Biosphere Reserve and were incorporated as a model to promote conservation in the face of traditional logging concessions, demands for petroleum exploration and an advancing agricultural frontier. Two of the communities studied in highlands Guatemala were in protected areas: one had established its own (informal) community reserve, the other was in a recently declared protected area; there are substantial pressures to establish protected areas in the remaining highland communal forests (Elías et al, 2009). The RESEX in Porto de Moz, Brazil, is a type of conservation unit managed by the state environmental agency. The Kalahan Education Foundation site in the Philippines is in the Kalahan Reserve. All but the Guatemalan Highlands sites have mixed conservation and livelihood goals.

Table 9.2 presents the changes in rights in practice across the cases. The cases are again grouped by the degree to which implementation was effective, and as expected, more effective implementation is associated with greater changes in rights. The table classifies changes in the bundle of rights for each reform. *Access* refers to access to the land and forest area and is classified as 'no change', 'increased' or 'consolidated'. *Consolidated* indicates more significant and secure long-term changes in rights. *Use* or *withdrawal rights* refers to forest resource rights, rated along the same continuum.

Management is more difficult to categorize because the state tends to control different resources to varying degrees and may set broad parameters while allowing a certain degree of local rule-making (see Chapter 3). In only two cases do local rules dominate, with almost no government interference; in Pando the state controls logging, but the primary livelihood activity is Brazil nut harvesting. The rest of the cases vary, with greater room for local rule-making in the cases that have both local rules and external control, and less in those classified as under external control alone. Exclusion rights are either weak or strong. In no case have alienation rights, which permit the sale or transfer of rights outside the group, been granted, although in most cases transfers within the group are permitted.

In general, all the cases with better implementation have strong exclusion rights and are much more likely to have not only increased, but also consolidated, use and withdrawal rights. Exclusion rights are often explicitly granted by the reforms, but if implementation is weak, as in Guarayos, exclusion in practice may be weak as well. Access is only seen as consolidated in three cases where land titles have been granted, in the Philippines ancestral domain case, the Trans-Amazon of Brazil and Pando in Bolivia, as well as in the community forests of Nepal (Guarayos presents a fourth case of land titling but with less effective implementation). In Nepal, forest lands have not been titled to communities but have been granted in perpetuity; the Forest Department can dismiss the Executive Committee of the Forest User Group on defined charges, but the forest cannot be taken back (Paudel, personal communication). Use rights are consolidated in most cases with consolidated access rights and several additional cases as well. Though the increase in rights in the Petén has been very significant in some ways, they are not noted as consolidated because of ongoing competing interests in the region.

With regard to management rights, it is difficult to find an identifiable pattern, in part because of the difficulties explained above and in Chapter 3.

Table 9.2 *Changes in rights in practice*

	Reform	*Access*	*Use or withdrawal*	*Management*	*Exclusion*	*Alienation*
More effective implementation						
Nepal	CBFM	Consolidated	Consolidated	External control	Strong	Not granted
Petén, Guatemala	Community concessions	Increased	Increased	External control	Strong	Not granted
Philippines (1)	Indigenous rights	Consolidated	Consolidated	Local rules and External control	Strong	Not granted
Rajastan, India	Tree planting	Increased	Consolidated	Local rules	Strong	Not granted
Pando, Bolivia	Agroextractive community	Consolidated	Consolidated	Local rules+	Strong	Not granted
RAAN, Nicaragua	Indigenous rights	Increased	Consolidated	Local rules and External control	Strong	Not granted
Trans-Amazon, Brazil	Colonization communities	Consolidated**	No change	External control	Strong	Not granted
Burkina Faso	Concessions,* communal forest	No change	Consolidated	Local rules and External control	Strong	Not granted
Less effective implementation						
Cameroon	Community forests	No change	Increased	Local rules and External control	Strong	Not granted
Highlands, Guatemala	Communal forests	No change	No change	Local rules and External control	Weak	Not granted
Philippines (2)	CBFM	Increased	Increased	External control	Strong	Not granted
Porto de Moz, Brazil	RESEX	No change	Consolidated	Local rules and External control	Weak	Not granted
Ghana (1)	Benefit sharing	—	—	—	—	—
Ghana (2)	Tree planting	Increased	Increased	External control	Strong	Not granted
Guarayos, Bolivia	Indigenous lands	Consolidated**	No change	Local rules and External control	Weak	Not granted

* One concession experienced only an increase in usufruct rights and weak exclusion rights.
** Access rights were consolidated for communities that have received title, but many others have not.
+ External control applies to logging, but the main livelihood activity in this region is Brazil nut extraction, which is not currently controlled.
Source: Elaborated based on country and site reports and discussions with researchers

It is important to note, however, that local rules may exist because the reform explicitly or implicitly supports them, or they may prevail simply because external control is too weak to suppress them. In all cases with effective implementation, local rules are explicitly permitted.

It is notable that some of the cases of poor implementation still demonstrate increases in some rights in practice. In the Philippines CBFM cases, rights have clearly increased even though the programme has moved forward in fits and starts. Communities in Cameroon have increased formal rights over a certain area of forest, although income benefits are often usurped by elites. In Porto de Moz, partial implementation has consolidated local use rights through expulsion of timber companies, but ongoing implementation has moved very slowly. Tree planting in Ghana has also increased rights but not particularly securely. In Guarayos, Bolivia, communities that have received titles have consolidated access rights, but many others are in a state of insecurity because of slow implementation and competition for the land and forests. Both the Philippines CBFM and tree planting in Ghana have strong exclusion rights. Because of the nature of the benefit-sharing scheme in Ghana, land-based tenure rights were not part of the reforms; hence, though this case potentially provides increased income from resource rents, the changes are not applicable to the table. With regard to management, local rules persist in the case of the highlands in Guatemala and the extractive reserve in Porto de Moz to some extent because the external controls are not effective; local rules are permitted in Guarayos for all activities except commercial logging.

In Table 9.3, the cases have been classified into three groups: those with little or no change in tenure rights, those with moderate changes and those with significant and secure changes.[2] This information is combined with the goal of the reform from Table 9.1, livelihoods or forest conservation and the outcomes. Unfortunately, it is not possible to assess quantitatively the comparative magnitude of changes across the cases; nevertheless, this will be discussed in qualitative terms when it contributes to the analysis.

It is now possible to analyse the outcomes with a greater understanding of the reforms and their implementation. To summarize, it is clear that the reforms so far have mainly focused on changes in access and use rights and that management rights are almost always subject to external control. Exclusion rights are important for effective implementation. Alienation rights have not been granted in any of the reforms (except rights internal to the community or group). Several cases have not been implemented effectively and this affects the extent to which rights increase in practice.

Tenure reform and change in livelihoods and income

As mentioned in the introduction, sources of livelihoods are changing throughout developing countries and many of the drivers are unrelated to forest resources. Nevertheless, forests remain important for livelihoods and incomes. Tenure reforms have sometimes made new forest resources available to communities; in other cases, however, communities may have had access that was informal or even illegal. In most cases, new rights were combined

Table 9.3 *Livelihood (L) and forest condition (F) outcomes, by goals of reform and changes in rights*

Change in rights	*Case*	*Reform goal*	*Change in livelihoods*		*Change in forest condition*	
Significant increase	Pando, Bolivia	L	Improved income from Brazil nuts in titled lands	+L	Maintenance of forest areas with limited pressures for conversion	=F
	CBFM, Nepal	F	Consolidated access to timber and NTFPs	+L	Increased forest cover, species diversity, fire control	+F
	Kalahan, Philippines	L, F	Some improvements from NTFPs and projects, but also use rights restrictions	+L	400 ha reforested, control of fires, sanctuaries established, rich biodiversity	+F
	TGCS, India	L, F	Small contribution to fodder and fuel wood	=L	Tree planting on highly degraded land, improved condition and diversity	+F
	RAAN, Nicaragua	L	Growing income from commercial logging only in some cases	+=L	Selective logging but no internal pressures for forest conversion*	=F
	Concession, Burkina Faso	L, F	Increased use of NTFPs, regulated use of fuel wood and fauna	+L	Deforestation due to market demand, population growth; other sites show recovery	=F
Moderate increase	Petén, Guatemala	L, F	Growing income from timber and NTFPs	+L	Selective logging but few internal pressures for forest conversion	=F
	CF, Cameroon	L	Growing community income derived from sale of forest products	+L	Degradation, deforestation and conversion to agriculture	-F
	Trans-Amazon, Brazil	L	More assets but little changes in cash income	=L	Converted and degraded because larger pressure from agriculture	-F
	CBFM, Philippines	L, F	Increased income from logging, agroforestry and coop enterprises, projects	+L	Reforestation, fire control, biodiversity improvements in most sites	+F
	Porto de Moz, Brazil	L, F	Consolidated access to NTFP but constraints to timber use	=L	Less logging, limited pressures for conversion but little change	=F
	Tree planting, Ghana	L, F	Promised future income from timber	=L	Increased tree cover	+F
Little to no increase	Highlands, Guatemala	F	No change	=L	No change	=F
	Benefit sharing, Ghana	L	Income to chiefs but not to communities	=L	n.d.	n,a
	Guarayos, Bolivia	L	Growing income from commercial logging	+=L	Selective logging and pressures for forest conversion	-F

+ Improvement; – deterioration; = no change; += small changes or changes explained in text

* The RAAN forest was badly damaged by Hurricane Felix in September 2007; this decline in forest condition is not taken into account here.

Note: The order of the cases within the three main categories is arbitrary.

Source: Elaborated based on country and site reports and discussions with researchers

with new responsibilities – and hence the reform may actually have placed new limits on access. This section discusses the results regarding livelihoods and income, identifying patterns.

In many of the cases, tenure reform has opened up new sources of goods for subsistence or income. For example, in the Petén, Guatemala and Cameroon, communities had no legal rights to timber or logging income prior to the creation of the community concessions or community forests. In India, communities were granted wasteland areas to grow trees for fuel and fodder. In Ghana, the Modified Taungya System for tree planting, unlike previous taungya programmes, allows farmers a share of income from the trees they plant.

In other cases, the reforms involved new restrictions on the use of resources previously available to the community. At times, open-access dynamics governed resource use, as in the Nepal cases and Kalahan, in the Philippines. But in the Petén, strong informal institutions governed access to some NTFPs. The new legal rights, then, both expanded and restricted access in some ways, with prior practices being brought under greater state control, monitoring and regulation. The most common restrictions cover grazing, logging and the use of fuel wood and fodder. Curiously, however, no cases present declines in livelihoods, for any of several reasons: the restrictions

1. were eased or forgotten with the passage of time;
2. affected only some members of the community or only outsiders;
3. were counterbalanced or outweighed by other benefits;[3] or
4. covered resources that the community had never used and had no interest in exploiting.

This last reason is the case of commercial timber, for example, in Kalahan and in many communal forests in the Guatemalan Highlands.

Some of the unexpected benefits of the reform are apparent at one site in Nepal:

> *The decade long violent conflict, economic stagnation, population growth and increased trend of going for overseas employment have significant impacts on the vulnerability of local poor. Although most of these trends negatively affected the poor, those who highly rely on forest resources have been less affected. The Community Forest User Group (CFUG) was the only functioning local institution during the political conflict and the states of emergency. The members continued to engage in managing, harvesting and sharing the benefits. Consequently, the benefits of forest management provided more stability during difficult times.*
> (Paudel and Banjade, 2008b, p41)

In that case, forest resources provided greater stability from outside shocks and the members of the user group, established by the tenure reform, were less vulnerable during a conflict. For other communities, the ability to exclude outsiders or choose whether to permit logging are important benefits of reform, even if this brings no direct livelihood improvement. This is true, for example, for some indigenous communities that find their territories the target of invasions by non-indigenous colonists and, of course, for communities that have long suffered from state logging concessions in their forests.

Table 9.3 suggests a possible pattern in the outcomes for livelihoods: reforms with significant increases in tenure rights generated positive livelihood outcomes in almost all the cases; in contrast, reforms with only moderate rights increases demonstrate no change in half the cases. The cases with the least rights changes more often resulted in no livelihood change, with the exception of Guarayos, where only a small minority experienced improvements, because they now engage in commercial logging.

Tenure security is expected to increase a community's long-term investment and provide an incentive for making and enforcing new internal rules (Gibson et al, 2000). In the case studies, it helped communities obtain project support, made it possible for them to sign contracts with outside parties (Larson and Mendoza-Lewis, 2009) and increased their bargaining power (Cronkleton and Pacheco, 2008b). For example, in Pando, Bolivia, some communities' rights were already relatively secure through customary institutions, but they were further strengthened against competing claims for access to Brazil nut trees. This increased the negotiating power of these communities and at least in some cases contributed to higher incomes. Guatemalan Highlands communities that had secure tenure (or were able to negotiate agreements with municipal governments for secure rights) could participate in a national incentive programme for natural forest protection and reforestation; for example, one study community in Chancol is earning an average annual income of US$366 per family from reforestation incentives (Larson et al, 2008).

Table 9.3 also suggests that the goals of the reform may affect livelihood outcomes. Of the cases that had moderate and significant changes in rights and a goal of improving livelihoods, all except one experienced livelihood improvements; half of the cases with combined goals did. There are too few cases with primarily conservation goals to be able to draw any real conclusions, but in the Nepal cases it is worth noting that livelihood improvements only emerged after grassroots organizations fought for this outcome. Other studies have shown that access to forests was reduced after launching community forestry, badly affecting those who were more forest dependent (Colfer et al, 2008a; Adhikari et al, 2004; Malla, 2000). But forest user groups and their federation (see Chapter 6) fought for policy changes. All the Nepal cases studies here enjoyed some level of livelihood benefits. The sites inside protected or conservation areas also show no definitive pattern, with no change in livelihoods in Porto de Moz, Brazil, and the highlands of Guatemala, but improvements in Kalahan, the Philippines, and the Petén, Guatemala.

The livelihood benefits must be put in perspective, however. Whereas some communities have benefited enormously from reforms, others in the same policy context may have fared less well because so many variables influence outcomes in every case. These include the forest management model, the quality of resources transferred to communities, the presence of project support, regulations and market conditions, local organization and the internal distribution of benefits.

The first variable, the forest management model associated with the tenure reform, can be better understood by looking at the magnitude of livelihood outcomes, specifically income: for example, a few communities began receiving large new sources of income after the tenure reform. Though the cases represent a range of different situations, two principal models of reform stand out. One is the community forest enterprise model, common to some Latin American cases, Cameroon and one site in the Philippines, whereby substantial external support, usually from donors and projects, helps establish a community-based logging operation. The other model is based primarily on support for subsistence needs or small-scale trade in NTFPs. Both may be driven by conservation objectives, though the former necessarily has significant livelihood goals. What the communities with substantially higher incomes have in common is the establishment of community logging enterprises.

Table 9.4 summarizes the collective profits, ranging from US$10,000 to more than US$200,000, in several of the communities studied. These profits represent the collective net income to the enterprise after costs, which are often substantial, and can be spent in different ways, such as for community projects or distributed as dividends among members. But these projects also provide employment and wage income. There is often a trade-off among these options. In the two Petén concessions, for example, Arbol Verde regularly distributed more than US$500 in annual dividends, but the community of Carmelita distributed only US$150 to US$250, investing the rest in creating jobs and hence increasing its operating costs (Monterosso and Barry, 2009). A comparison of the four enterprises in the Latin American sites (Carmelita and Arbol Verde in the Petén, Layasiksa in Nicaragua, Cururú in Guarayos, Bolivia) demonstrates investments of US$22,000 to US$43,000 in wages and US$6000 to US$33,000 in the community – in school scholarships, community

Table 9.4 *Profits from community forestry enterprises*

Site	*Net community income US$*
Layasiksa, Nicaragua	30,264
Arbol Verde, Petén, Guatemala	226,315
Carmelita, Petén, Guatemala	27,745
Compostela, Philippines	23,400
Cururú, Guarayos, Bolivia	34,486
Lomié-Dja, Cameroon	10,002

Sources: 2006 and 2007 data from Pulhin and Ramirez (2008), Larson et al (2008) and Oyono et al (2008)

water systems, the construction of housing for the poorest members and so on (Larson et al, 2008).

These enterprises, however, operate in only some of the sites studied. The outcomes were significantly different from those of neighbouring communities under the same tenure reform but without enterprises. For example, the second site studied in the RAAN, Nicaragua, demonstrated no measurable livelihood or income improvements directly associated with the tenure reform; the same is true in other Philippines sites without this enterprise model. In both Guarayos and Cameroon, other communities had enterprises as well but with much more modest profits of about US$3200 in Guarayos (Larson et al, 2008) and from US$3750 to US$6040 in four other sites in Cameroon (Oyono et al, 2008).

Data demonstrating high incomes, however, do not mean that these models are necessarily better. Many enterprise models involve substantial donor or project support and outside investments; they often result in significant community upheaval and the transformation of local traditions and institutions, for better or worse; they involve high financial costs and risks, may create permanent external dependency and are difficult to replicate (see Larson et al, 2008; Pacheco et al, 2008b). In Cameroon, funding often comes from members of the local elite, who then confiscate all financial benefits (Oyono et al, 2008). However, the outcomes do suggest what can be achieved in some cases – though this also depends on the quantity and quality of forests, as discussed below.

The second model – a collective traditional model for domestic use or small-scale trade – has dominated reforms in Nepal and Burkina Faso and is similar to the tree growing reform in India. Though most of these have resulted in livelihood improvements, the magnitude tends to be much smaller and may not include income at all. One factor is the dramatic difference in scale

Table 9.5 *Changes in livelihoods, by management model*

Change in livelihoods	*Collective traditional*	*Collective entrepreneurial*	*Individual*
Relatively larger +L		Petén, Guatemala CF, Cameroon CBFM, Philippines RAAN, Nicaragua* Guarayos, Bolivia*	
Relatively smaller +L	CBFM, Nepal KEF, Philippines Burkina Faso		Pando, Bolivia
=L	India Highlands, Guatemala		Ghana Porto de Moz, Brazil Trans-Amazon, Brazil

* Communities with entrepreneurial models only
\+ Improvement; – deterioration; = no change

between the newly tenured forest areas in Asia and Africa, on the one hand, and Latin America on the other. One community forest user group in Nepal reported an income of US$3350 for the collective and a total household income contribution of US$2960 (Banjade and Paudel, 2008). This is considered quite high among community forestry sites in Nepal.

Again, smaller income benefits do not mean the reform is necessarily less desirable. Strengthening and supporting appropriate and sustainable agriculture or small-scale NTFP trade can still improve livelihoods and may be particularly important for promoting women's opportunities and family health as well as cultural diversity (Colfer et al, 2008b; Colfer and Byron, 2001). Nevertheless, there is good reason to believe that, in some cases and when desired by the community, the traditional collective reform model has greater potential to contribute to people's livelihoods and incomes than it does now. For example, in Nepal, environmental concerns have been dominant because this model of community forestry was originally promoted to halt rapid deforestation and protect and conserve forests (Kanel, 2004; Kanel et al, 2005; Sunderlin et al, 2005). Livelihood and poverty alleviation objectives emerged as second-generation issues over the years because of grassroots demands.

Conservation priorities can have repercussions regardless of the model. In the Philippines, for example, community-based forest management policies consider livelihoods and conservation of equal importance, but this commitment has not been demonstrated in practice. Rather, conservation 'fears' have led to constant policy reversals and the periodic cancellation of all resource use permits; because of this, in 2003, the enterprise in Compostela was left with almost US$56,000 in debts. The problems with regulation have been discussed in Chapters 3 and 7.

A third set of cases are grouped as 'individual models' in Table 9.5. Among them, only Pando experienced livelihood improvements. In Ghana, the benefits from tree planting will not be forthcoming until the trees have matured and no provisions have been made for farmers to borrow against future income. In Porto de Moz, although communities have been able to exclude timber companies, the benefits that they obtain from the forests have not improved significantly. In the Trans-Amazon, the reform only formalized the access that communities already enjoyed and so it did not have any significant implications for livelihoods.

Another central issue shaping the livelihood potential of forests is the quantity and quality of the forestland assigned to communities, which also influences the choice of model. Community forests are rarely located in high-quality forests. Community forests in Cameroon are granted from the lower-quality forests of the non-permanent estate, equivalent to the agroforestry zone near villages (Oyono et al, 2008). Very few high-quality *terai* forests have been granted to user groups in Nepal (see Ojha et al, 2008). Rather, the forests handed over to communities for protection and management have been degraded, sometimes heavily, particularly in Asia. These forests provide little prospect of generating income until they are replanted and fully regenerated.

The community concessions of the Petén – though the quality varies – appear to be an exception. In addition, the size of the forests granted to communities in the Petén and many other Latin American sites is orders of magnitude larger than in Asia, in particular. In Nepal, for example, the sites range from 100 to 635 hectares, sometimes less than 1ha per member, whereas in the Petén, one of the concessions studied covers 65,000ha, or 190ha per member; titling in Pando granted 500ha to each family to promote sustainable Brazil nut extraction.

Another factor affecting livelihoods across the cases is the extent to which the reform includes some kind of project support. In all but one of the cases demonstrating improvements in Table 9.3, the reform did not simply change tenure rights but also provided economic, technical and organizational support. Though this is essential in all of the sites with logging enterprises, mentioned above, it also includes the community forest user groups in Nepal, the ancestral domain site in the Philippines and the concessions in Burkina Faso. Sites that received project support but did not demonstrate significant livelihood or income changes were the two involving tree planting (India and Ghana). The only site that demonstrated income improvements but did not have project support is Pando, Bolivia, where the cooperative federation has received support to organize and gain access to fair trade markets (see Chapter 6). Such support has proved important for building community capacity, navigating the national bureaucracy and accessing markets, all of which affect outcomes. These issues (community organization, regulations and markets) have been addressed in other chapters and will not be repeated here.

Tenure reform and change in forest condition

Patterns in forest outcomes are even more difficult to discern than patterns in livelihoods. The results (see Table 9.3, above) demonstrate only slightly better outcomes in the cases with significant increases in rights than those with moderate increases. The 12 cases with significant or moderate increases in rights show more forest improvement than the two cases with little or no increase in rights. Forest condition improved in half the cases with significant gains in rights and saw no change in the other half. Of the cases with a moderate increase in rights, outcomes were evenly divided between improvement, deterioration and no change. The two cases with little or no increase in rights resulted in no change or a decline in forest condition.

As with livelihood outcomes, changes in forest condition appear to be related to multiple variables, making it difficult to isolate the effect of the tenure reform. Nevertheless, we have discerned some patterns and a closer look at outcomes by case offers insights into the other variables likely to be relevant.

The explicitness of conservation goals does not appear to make a difference in outcomes (see Table 9.6). Though not all of the cases with forest conservation goals had positive outcomes for forests, none experienced declines in forest condition. On the other hand, since these indicators summarize several sites,

Table 9.6 *Forest outcomes based on forest and livelihood goals*

Reform goal	*Outcomes for forests*		
	+F	*=F*	*-F*
Livelihoods		Pando, Bolivia RAAN, Nicaragua	CF, Cameroon Trans-Amazon, Brazil Guarayos, Bolivia*
Livelihoods and forest conservation	CBFM, Philippines TGCS, India Kalahan, Philippines (C) Tree planting, Ghana	Porto de Moz, Brazil (C) Petén, Guatemala (C) Concessions, Burkina Faso	
Forest conservation	CBFM, Nepal	Highlands, Guatemala* (C)	

+ Improvement; – deterioration; = no change; += small changes or changes explained in text
C: Formal conservation areas; this applies to the Guatemalan Highlands only in some communities
* Sites with little to no increase in rights

interpretation is not always straightforward. Of the four sites in Burkina Faso, for example, two experienced declines and two experienced improvements; the differences between these sites are discussed below.

In the five cases in which forest condition improved, the results are not particularly surprising. Two involve tree planting initiatives, which would result in negative outcomes only if no trees were actually planted. The community forestry projects in both Nepal and the Philippines have been faulted for an overemphasis on forest condition and insufficient concern for local people's livelihoods. The Kalahan site was an open-access area until the communities won formal rights and control.

Declines in forest condition are seen only in sites where livelihoods or user rights were significantly more important in the reforms than conservation goals; nevertheless, conservation goals were not ignored in these sites. For example, community forests in Cameroon and communities undertaking logging in Guarayos, Bolivia, all operate with approved forest management plans.

The establishment of conservation or protected areas does not have a direct relation to positive outcomes for forest condition. Those sites are noted (C) in the table; only one has resulted in improvements; three have seen no change. Again, the results suggest that other variables may be more important.

World region provides one of the most notable patterns in outcomes. Results were much more likely to be positive for forests in Asia, be mixed in Africa and result in no change in Latin America (see Table 9.7). Each region is examined in turn to identify the underlying variables behind these differences.

Forest condition clearly improved in almost all sites in Nepal, India and the Philippines. Under reformed tenure, forest cover has increased, natural regeneration has been protected, landslides have been reduced and some of the endangered flora and fauna have been safeguarded. A significant reason is that most of these forests were highly degraded when handed over to communities. In the midhills of Nepal, the condition of forest cover dramatically improved, particularly in terms of increased canopy cover and basal area. In the forests of Sundari forest user group in Nawalparasi (in the high-value lowland *terai* forests), there was significant regeneration, even with relatively high levels of timber extraction.

All sites in Nepal experienced increased availability of fodder, fuel wood, leaves, NTFPs and timber. Seasonal availability expanded, time required to collect these products decreased and the net quantity collected increased. These general observations were confirmed by more rigorous technical assessments from user groups' operational plans. For example, fuel wood biomass (kg) per ha in the Patle user group rose from 75 to 103 cubic metres from 2002 to 2007. In Nawalparasi, biomass increased from 61 to 115 cubic metres in the same period. Nepal appears to be an exception in this regard and, as noted previously, is one of the only cases in which forest conservation was clearly a priority over livelihood goals; it was not until forests had substantially regenerated that community access and withdrawal rights increased. Nepal was also one of the first countries to develop community forestry policies and thus its programmes have a longer history and greater maturity.

In the Philippines, one factor leading to improved forest condition was the effort made both by the state and the communities to reforest denuded areas. For example, the Kalahan community reforested more than 400ha in its own forest reserve, protecting the watershed and biodiversity and reducing wildfires. In the community-based forest management sites, the results have been more mixed, with two sites experiencing reforestation, control of wildfires and overall improvements, but one declining in overall condition despite reforestation, because of illegal poaching and logging. This site, Compostela, generated significant income from forest enterprises (mentioned above), experienced the effects of permit suspensions most strongly and is subject to overlapping claims between indigenous communities and more recent migrants; all of these factors affect tenure security. Worsening forest condition appears to be associated

Table 9.7 *Outcomes for forests, by world region*

Change in forest condition	*Africa*	*Asia*	*Latin America*
+F	Ghana	Nepal Kalahan, Philippines CBFM, Philippines India	
=F	Burkina Faso		Petén, Guatemala RAAN, Nicaragua Pando, Bolivia Porto de Moz, Brazil
-F	Cameroon		Trans-Amazon, Brazil

+ Improvement; – deterioration; = no change

not with logging conducted under the resource use permit but rather with the failure to exclude illegal loggers and poachers.

In India, all three sites have seen a positive local ecological impact from the tree growers' cooperative societies. Each cooperative has raised plantations on approximately 40ha of leased land. In all three cases, the cooperatives were able to control illegal encroachments before planting. Considerable effort and funds (through project assistance) were invested in preparing the site, building soil and moisture conservation infrastructure, establishing the plantation, watering tree saplings with water tankers and protecting the site against illicit grazing and removal of tree products. Ten years since external support ended, plantations in all three sites are still present.

In contrast with the Asia sites, all the sites in Cameroon suffered from some degree of forest degradation, though in one site, Oku, overall conditions were improving. In Burkina Faso, conditions were declining in two sites and improving in two others. Variation across the sites provides important insights.

In Cameroon, deteriorating forest conditions may be partially a result of the reforms. Though degradation was already occurring, there is a lack of environmental concern at the local level and failure to implement management plans appropriately, combined with a lack of monitoring from the Ministry of Environment and Forests. At the same time, the nature of the reform itself has been highly problematic. The process, which requires extensive and expensive bureaucratic procedures for obtaining community forests and for the periodic approval of management plans, has been fraught with corruption and captured by the elites who provide the funds. These include local and external elites, business people, top military officials and town-based politicians, whose primary goal after the long approval process is to recover their investment and make a profit (Oyono, personal communication).

The only site in Cameroon that demonstrates improved forest condition (Oku) is the one that is more traditional and hierarchical, where customary rules for conservation and resource use, as well as the 'mystique of social order', have been maintained under the influence of powerful chiefs (Oyono et al, 2008). The three other community forests demonstrate not robust collective action but the usurpation of the forest management committee by a small, unaccountable group of elites. To some extent, the involvement of these elites in the management of community forests is leading to the degradation of forests (Oyono, 2005a). Also, some community members believe that the community forest programme, while increasing rights to a small area, in some ways reduces their rights as a whole by recognizing formal rights only to an area much smaller than the one they have customarily claimed. For example, in Oku, 'farmers feel that their forest, which had measured about 17,000 hectares 10 years ago, has officially been reduced to 2800 hectares today' (Oyono et al, 2008).

The issues in Burkina Faso are more complicated. Two sites show improvements. One is a communal forest subject to overuse and degradation, and the neighbouring communities sought project support specifically for forest regeneration for future exploitation. The other is a wildlife reserve that generates hunting (safari) royalties; conservation is a priority here because wildlife habitat represents income. In both cases, the communities are well organized, customary authorities are fully involved in implementation and exclusion rights are exercised. Burkina Faso also has two cases of increased degradation that involve concessions for fuel wood exploitation. Although forest management plans exist, the provisions protecting forest resources are not implemented. In addition, customary authorities sometimes take actions to undermine the concessions, such as granting farmland to migrants inside the forest management area. Hence exclusion rights are not fully exercised.

There is a fundamental contradiction between the state's claims to own and manage land and forest resources and customary rights and practices. Policies have undergone significant changes several times since the 1980s (see Chapter 4). The communities that have received forest rights through concessions have not always been granted the rights that they have customarily claimed. Village commissions established by the government to manage natural resources initially excluded customary authorities and new village development councils, elected by communities through consensus or election, have been instated only since late 2007. The overall classification of forest regimes (zoning) by the state creates overlapping rights between central government and local government, industry, traders and local communities, encouraging exploitation by traders and elites at the cost of the social good. Official state-managed areas may undermine forest protection by customary institutions (Kante, 2008).

The two sites in the third African country, Ghana, demonstrate improvements in forest condition. The Adwenase forest, historically managed under exclusive control of the people of Akropong community as a sacred grove and royal burial ground, was being threatened by migrant settlers and by conversion to other uses by the community itself. The community sought the support of the Forestry Commission to save it and it is now managed as a 'dedicated community forest reserve'. The other site is located in a protected area and involves tree planting in agricultural fields under the Modified Taungya System, mentioned earlier, whereby farmers will then have a right to the income generated from their sale. Planted areas over the past three years have exceeded goals in two of those, resulting in about 3000ha planted from 2006 to 2008.

The Latin American sites generally saw no change in forest conditions, for various reasons. To begin with, compared with the Asia cases, the forests were in reasonably good condition when granted to communities. This is particularly true in Pando, Bolivia, where Brazil nut collection is the primary source of livelihoods, thus creating an economic incentive for forest conservation. Forests are, therefore, at less risk from degradation, since there is no evidence suggesting overexploitation of Brazil nuts. Tenure reform, the titling of community lands, had little impact on forest condition in part because forests are already fairly well protected.

In many of the other sites – in Brazil; in Guarayos, Bolivia; in the Petén, Guatemala; in Nicaragua – pressures from logging are increasing and the demand for land by colonists is high. Communities near roads and populated areas are more vulnerable and in general are suffering greater deforestation and degradation than more remote communities, which tend to have better-preserved forests and fewer people. It is likely that secure tenure alone in vulnerable areas – places where livelihoods depend on agriculture and population growth rates and colonization pressures are high – will be insufficient.

A comparison of Porto de Moz (Brazil), Guarayos (Bolivia) and the RAAN (Nicaragua) also offers some insights into pressures on forests under tenure reform. All three sites have suffered from serious delays in the implementation of rights due to foot dragging or bureaucratic weaknesses. In the first case, this involves the development of the management plan for the reserve, the first step before further implementation can advance, including the definition of resource use rights; in the other two cases, this involves demarcation and titling of indigenous territories. Yet deforestation does not appear to be related as much to tenure rights as to the location of multiple actors who are making demands on forests and forest resources. Forest conditions in Porto de Moz have remained the same. In large parts of the RAAN, they have also remained the same, though more vulnerable areas subject to colonization by *mestizo* farmers have been systematically deforested (Intelsig, 2008). Similarly, deforestation in Guarayos has occurred in areas exposed to pressures from large landholders to convert forest to mechanized agriculture; also, forests are being degraded in communities engaging in informal logging and agriculture. At the same time, one of the first few certified community forests is in Guarayos.

Forest condition in the Petén sites is good, with SmartWood studies showing more than 150 species of mammals and 300 species of birds. Both sites cut only large trees, more than 55 to 60cm in diameter, and often log far less than the permitted three cubic metres volume per hectare. Deforestation data from three management models – the buffer zone, the multiple-use zone, which is home to the community concessions, and the national park nucleus zones – show lower rates every year for the multiple-use zone from the period 1990–1993 to 2004–2005 (Monterroso and Barry, 2009). That is, the community forest concessions have much lower deforestation rates than unmanaged portions of the Maya Biosphere Reserve and other national parks, which are being invaded and converted to other uses. At the same time, the four small, vulnerable concessions on the edge of colonization areas have higher deforestation rates than the others.

In summary, the primary variables affecting forest condition outcomes across the sites are the nature and priorities of the reform (such as strict conservation rules and reforms primarily involving tree planting), the resulting security of rights, the maintenance or breakdown of customary or traditional management institutions, elite capture of benefits, dependence on agro-extractive activities, proximity to colonization areas or other competing interests in forests and the capacity of community forest management organizations.

Tenure reform and issues of equity

In some of the reforms, the granting of rights to a certain group of people excluded, failed to take into account or altered the rights of other groups. This problem most commonly affects people who use resources on a temporary or seasonal basis. In Nepal, for example, transhumant pastoralists in high mountain areas have traditionally used certain pastures seasonally. Some of those pastures have now been granted to communities under the community forestry programme and since grazing is seen as environmentally 'bad', many user groups, which have exclusion rights, have banned it.[4] After several years of conflict in the study community of Suspa-Dolakha, an agreement was finally reached with *chuari* herders to permit grazing at higher elevations. This has forced more herders into smaller areas, however, thus increasing pressure on natural resources; the population of herders in this area has dropped, from 35 to 40 prior to the establishment of the community forest to 16 today. Transhumant pastoralism contributes to the economy of Nepal and is common to the high-hill ethnic groups like the Sherpa, Bhote and Tamang. The problems these pastoralists face are poorly understood and generally ignored by policy-makers (Banjade and Paudel, 2008).

A similar problem is found in Cameroon with the Pygmy population. In villages composed of two ethnic groups, the dominant Bantu population denies the Pygmies' historical customary rights to forests, leading to their *de facto* exclusion. A Pygmy from Mintoum stated, 'The Bantu say that we are nomads, without fixed residence and village. They say that it is they who created the village, without us, and that the forest therefore belongs to them' (Oyono et al, 2008). Similarly, in the case of Adwenase, Ghana, only the rights of native community members were respected, while the rights of migrants, though recognized in the forest management plan, were prohibited in practice (Marfo, 2009). Land claims based on indigenous rights may deny the claims of non-indigenous people, as in the Philippines.

Within the communities benefiting from reforms, equity refers to participation in decision-making and in the distribution and sharing of material benefits and burdens associated with the tenure reform. Two issues stand out: first, tenure reforms sometimes place new restrictions on resource use that affect the poor most; second, power and benefits tend to be concentrated among certain community groups, even when there is not elite capture, unless specific measures are taken to reduce inequities or address the needs of poorer groups. Some communities have begun to implement such measures.

Several cases demonstrate restrictions on resource use that affect the poor. In Nepal, for example, forest use was highly restricted in the early years after the forest user groups were created. This included restrictions on the use of fodder, regulations and even prohibitions on grazing, and severe restrictions on gathering firewood or producing charcoal. These rules affected the poorest, most forest-dependent groups most, though in some sites these restrictions have been reduced with forest regeneration. Nevertheless, in Nepal, access of the poor often decreased as the commercial value of the products increased. For example, forest user groups began to restrict individual appropriation of NTFPs (*lotka* and *argeli* in Suspa, and *harro, barro, amala* and *kurilo* in Sundari) as their market prices rose. Instead, villagers were permitted to collect these on a wage basis for the user group to then sell in the market. The same has happened for fuel wood in Sundari: free collection is restricted to two seasons and during the rest of the year fuel wood must now be purchased.

In the Guatemalan Highlands, the state and conservation NGOs have promoted several resource use rules that have affected the livelihood options of poor people. For example, prohibitions and limits on sheep grazing, which has been blamed for the destruction of highland forests, have mostly affected women, who are the primary herders. Similarly, rules requiring permits for the use of firewood have hurt the rural poor. Though the permits themselves are not expensive, the time and effort spent obtaining them can be substantial (Elías et al, 2009).

Elite capture and problems with representatives who are not downwardly accountable promote the concentration of benefits in the hands of a few community members (see Chapter 5). In Cameroon, income can be traced to improvements in basic community infrastructure and social services in only one site. But even without specific accountability problems, power and benefits are likely to be concentrated, to some degree, among the male, better-educated, higher-caste or wealthier residents. From the start, rules for group membership may allow only one member per household, thus often excluding women. In Layasiksa, a change in the rules increased the number of women cooperative members from 14 per cent to 50 per cent. Nevertheless, there is only one woman on the board of directors and women's employment as wage labourers in the community enterprises is very low. This is a common problem across all the cases where timber is the primary product. In Layasiksa, for example, in all of the jobs created in the logging enterprise throughout the year, only two women participate, as cooks.[5]

Non-timber forest products can create opportunities for women. In the Petén, Guatemala, the expansion of the concession organizations into NTFPs has opened more spaces for women, where they are more engaged in tourism and export of *xate* palm. This has also opened up leadership possibilities and a woman held the position of cooperative vice president in Carmelita at the time of this study.

Finding fair ways to distribute limited new opportunities is a challenge for communities but helps prevent divisions and conflict. The logging project in Layasiksa has a rotational labour system whereby every willing and able male can participate for a certain number of days, as long as his work performance is acceptable. Skilled jobs are limited, but people can ask to be considered and trained.

Nepal's forest user groups are dominated by well-off members of the community who have generally benefited more than the poor. Women's participation is often token. At the same time, forest user groups in all four sites have taken some pro-poor initiatives, such as preferential prices and jobs, special income-generating projects and land allocation, as described in Table 9.8.

Table 9.8 *Pro-poor initiatives under community forest management in Nepal*

Initiative	*Description*
Pro-poor income generation programme	Poor households received financial and technical support to run small enterprises (goat raising, Kurilo farming, beekeeping, Machino farming). Hundreds of poor are benefiting.
Land allocation to poor households	Small pieces of land adjacent to forest have been allocated to poor farmers to grow commercial fodder, fruits and NTFPs (Baglung and Nawalparasi).
House construction for the landless	Sundari has built small houses for some poor who did not have permanent shelter. Six houses have been constructed and others are planned.
Differential prices for rich and poor	In some groups, timber is given free to poor during emergencies. Sundari has differential price system, with poor people paying less for timber, despite strong resistance from better-off members. One cubic foot of good-quality timber costs 325 rupees for rich, 275 for middle class and 225 for poor.
Priority for forest management jobs	More than 30% of user group's income goes to forest management and timber-harvesting activities. Poor members often get priority in these jobs.

In the Philippines, community forest management was introduced to redress inequities arising from previous concessionary regimes whereby outsiders benefited from forests and the rights of forest dwellers were denied. Because of this, tenure reform is seen as an example of social justice and a way to promote equity. With their new forest rights, the Banila community crafted new policies to reduce existing inequities by developing special provisions for the poor and disadvantaged members in the community; in Barobbob and Kalahan, the community forest committee established norms to ensure equal participation of women and men, rich and poor. Across the study sites, villagers were asked to assess changes regarding distribution of rights among members, participation in decision-making and community forestry activities, access to livelihood opportunities, sharing of income and benefits, sharing of costs and responsibilities and access to leadership roles. In general, equity was perceived to have improved across all these dimensions as a result of tenure reforms.

In the case of the tree grower cooperatives in India, protecting common lands from outside encroachers is considered important for equity, since poor people often depend on common lands for their survival. Nevertheless, many households in the community had not joined the cooperative and thus were not formally entitled to claim benefits or dividends from its activities. Only Khumariya gave special consideration for non-members who were unable to pay their share to join. At the same time, many people could not remember whether they were members, and all families in the community were actually given equal rights regardless of formal membership. All the cooperatives were dominated by men (with no women members registered) and by higher-caste groups.

In the Africa cases, men are dominant as well. In Cameroon, women are excluded completely from some management committees but are increasingly active in others. Interestingly, given the sharp intergenerational conflict common in Cameroon, the representation of youth in local organizations and forest management committees has increased over the years, to the point that in some cases they are the dominant members, respected by the elders. In Burkina Faso, women are actively participating in forest management activities, but they are ignored in the selection of representatives for local committees. The situation is similar in Afram Headwaters in Ghana, where women are involved as family heads in the Modified Taungya System only when men are absent, primarily because of the male bias in family matters. For example, even among matrilineal groups, husbands are considered the family head and representative (Marfo, personal communication).

In conclusion, equity is a complex issue with multiple dimensions, but there is little indication that increased tenure rights alone have had a positive effect. In several cases, securing rights to one group involved ignoring the rights of others and in other cases community members themselves defined the community to exclude certain groups. Rights associated with substantial responsibilities or resource use restrictions may adversely affect poor populations, the groups that depend most on forest resources for livelihoods. It is notable that the most significant attempts to take poor people's interests into account were in Nepal, a country undergoing massive political upheaval and a place where a powerful discourse of inclusion and overcoming traditional inequities has taken hold. Efforts to include women show only very slow progress.

Conclusion

A significant improvement in legal tenure rights does not automatically result in improvements in livelihoods, forest condition or equity. At the same time, in our cases, results for both livelihoods and forest condition were better for cases with a significant increase in rights and declined as the increase in rights declined. There were some trade-offs, however. Several cases involved times of hardship, or livelihood declines for certain members of the community or groups of people external to the community, while forest condition improved; in some cases rights entailed substantial responsibilities and burdens. Conversely, livelihood improvements were sometimes associated with declining forest condition. Perhaps most notable from a rights perspective, however, is that livelihoods improved in a number of sites *without* declines in forest condition (see Table 9.9).

It is important to analyse each case in context to understand those outcomes. With regard to livelihoods and income, three major mediating variables affect results:

1 the quantity and quality of forest resources granted to communities;
2 national regulations (including the conservation-related limitations established by the reform); and
3 market conditions and forms of market engagement.

Table 9.9 *Synergies and trade-offs between changes in livelihoods and changes in forest condition*

Changes in livelihoods	*Changes in forest condition*		
	+F	*=F*	*-F*
Relatively larger +L	CBFM, Philippines*	Petén, Guatemala RAAN, Nicaragua (some sites)	Cameroon Guarayos, Bolivia (some sites)
Relatively smaller +L or +=L	Nepal KEF, Philippines	Pando, Bolivia Burkina Faso RAAN, Nicaragua (some sites)	*Guarayos, Bolivia (some sites)*
=L	India Ghana	*Porto de Moz* *Highlands, Guatemala*	Trans-Amazon

+ Improvement; – deterioration; = no change; += small changes or changes explained in text
Italics: Cases classified previously as having little or no increase in rights
* The indicators represent composites; the site with the larger income benefits was one in which forest condition was declining (because of outside encroachment).

Most important for livelihoods, reforms need to be fully implemented, with follow-through that facilitates the communities' ability to obtain benefits from forests.

Change in forest condition is affected by these factors:

- the starting condition of the forest and the extent to which the reform prioritizes conservation or regeneration;
- dependence on agro-extractive activities, which generates an economic incentive to conserve the forest, and/or a culture strongly linked to forest maintenance;
- proximity to colonization areas or other competing interests in forests, including population growth, industry and market demands, that are beyond the control of communities.

Central to protecting forest condition is a community's right and ability to exclude outsiders, especially logging companies and those who would convert the forest to other uses.

With regard to equity, the central finding is that positive outcomes appear to depend on specific, dedicated efforts to address sources of inequity. Such efforts should thus be built into future reforms.

Notes

1. Special thanks go to Bocar Kante, Phil René Oyono, Naya Sharma Paudel, Juan Pulhin and Emmanuel Marfo for extensive time and effort spent offering clarifications on case-level and country-level outcomes.

2. All cases with consolidated rights to access or withdrawal and strong exclusion rights were classified as having a 'significant increase' in rights, except for the case where only a limited number of communities had received title (Trans-Amazon). This case and all cases with increases in access and use rights with strong exclusion rights were classified as 'moderate increase'. The RESEX was also classified as 'moderate' because of consolidated use rights, though exclusion is still weak. The remaining cases were classified as having little to no increase in rights; again, Guarayos is not straightforward because those in more remote areas have received titles and consolidated access rights, but this is not the case in the more populated areas. The Ghana benefit-sharing scheme is included in this group because in practice these benefits have not reached the community level.
3. A final reason is probably community adaptation and optimism.
4. Natural pastures at high altitudes constitute about 78 per cent of Nepal's pastureland (Banjade and Paudel, 2008).
5. See Colfer (2005) for more on women and sustainable forest management.

10
Conclusions and Reflections for the Future of Forest Tenure Reform

Anne M. Larson, Deborah Barry and Ganga Ram Dahal

With contributions by Carol J. Pierce Colfer, Peter Cronkleton, Emmanuel Marfo, Pablo Pacheco, Naya S. Paudel and Juan M. Pulhin

This book has explored the experiences of forest tenure reforms in 11 countries, across dozens of regions and communities, with the goal of understanding their origins, processes of implementation and outcomes for local life and forest conditions. As we have seen, these reforms range from those that are somewhat older to those that are incipient and vary from new revenue rights and short-term concessions to full-fledged statutory ownership and land titles. The granting of rights has sometimes transferred limited new rights or taken away others and has often been laden with responsibilities to conserve forests, but it has also offered new livelihood opportunities and/or improved forest condition in many cases.

These tenure reforms cannot be fully understood without knowledge of the political-historical context of each country and the dynamics of other important processes affecting governance at the same time. Their outcomes cannot be separated from the many social processes in which they are embedded. This book, however, has focused on the reforms as a little known or understood global trend – and has thus sought to understand both its breadth, across nations and world regions, and the in-depth issues it involves.

To summarize the vast set of experiences and issues, this concluding chapter first reviews some of the principal findings and then discusses central issues and concerns raised by the reforms. This is followed by a discussion of the emerging challenges of global climate change in light of the research. The chapter closes with a short proposal for the future of tenure reform.

Research findings

Forest tenure reforms have arisen for a number of reasons. 'Top-down' reforms have been developed because of concern over deforestation, to share conservation costs, to obtain support for government policies, to promote social justice and rights under new democratic regimes, to respond to donor pressure for larger reforms and to appease internal dissent or demands. 'Bottom-up' reforms have emerged because people see opportunities to reclaim historical rights to forests that have been taken away, or because the forests over which they have customary rights are being invaded or threatened by outsiders. At times, reforms have arisen when communities seek help from the state for forest management or conservation.

Taken as a whole, forest tenure reforms are different from past land or agrarian reforms in that rights are granted over collective, rather than individual, properties and alienation rights, or the right to sell the land, are not granted. In addition, the state maintains an important management role in relation to the expectation – or rule – that forests remain intact. Land is not redistributed; rather, rights tend to be granted to people already living in and using forests. Finally, reforms are aimed not only at livelihoods or development concerns (and sometimes land rights), as in the past, but also at addressing ancestral rights of indigenous communities and promoting forest conservation.

Indigenous rights movements have been a major driver of reform, particularly in Latin America; in Africa, in part because of overwhelming formal state ownership of forests, decentralization has been the principal driver, though tenure reform was not necessarily among its goals. Both of these forces have played some role in Asia, as have community forestry policies in some countries. Global conservation interests and actors have shaped the nature and extent of reforms in all three regions.

Indigenous demands have been central in the introduction of rights-based approaches to reform and may have achieved the most in terms of the extent of reforms – but they also may have met the most resistance. Decentralization has provided opportunities for greater local decision-making but is faced with the challenging interface between statutory change and customary practices and authorities, as well as the ongoing tendency of the postcolonial state to centralize power. Conservation interests have guaranteed that attention to forest conservation is taken into account in reforms but often at the expense of community rights and livelihoods, and possibly even of customary practices that have sustained forests as well.

Forest reforms – at least those that have been effectively implemented – have generally granted use rights and exclusion rights to forests, but management rights have involved varied and sometimes complex combinations of local and state decisions and responsibilities. In some cases, the state retains all the decision-making power and communities are left only to implement responsibilities to protect forests, but often the balance of power is more complex. For example, there is usually a distinction between higher-value (often timber) and lower-value (often non-timber) resources, and/or between commercial and subsistence uses, with communities granted greater decision-making over the latter and less over the former in both cases. Management rules, which are often mandated by the state, also place restrictions on withdrawal rights. Hence the bundle of rights is not cumulative in practice, as it is commonly conceived in theory. Rather, if exclusion rights are granted, the nature of management rights may be one of the deciding factors that characterize the *extent* of the reform, as it defines the degree and nature of decision-making that is permitted in the local arena. In general, retention of major management rights by the state has attenuated reforms and the recognition of local rights. Another issue is the permanence of the reform: whether it is temporary, revocable or granted in perpetuity. For example, rights may be granted to communities through presidential decree, forest acts or regulations, but all of these are vulnerable to unilateral reform. Rights granted through laws are more secure, and a constitution even more so.

The granting of certain rights through reform may actually have the effect of taking rights away where communities are already living in forests and already have local institutions for land and forest access and management. These institutions may be based on customary or other *de facto* rights. This clash between statutory and customary systems is most apparent, and has been most studied, in Africa but is relevant to some degree in most sites that have some level of functioning collective action or institutions. The state may seek to suppress, ignore or support these local or customary institutions (Benjamin, 2008), though the effect of ignoring them may also be suppression. For example, the granting of rigid exclusion rights to sedentary communities is often done without consideration of the customary rights of seasonal resource users, such as transhumant pastoralists.

The imposition of state rules and interests over existing customary practices is likely to result in 'sterile dualisms', whereby 'impracticable state law [coexists with] unauthorized local practices' (Benjamin, 2008, p2256), or 'forum shopping' (von Benda-Beckmann, 1981), in which people choose which rule they will follow based on their particular interest. It may also undermine effective local institutions and lead to open-access dynamics (Fitzpatrick, 2006). At the same time, not all local institutions are effective at forest management, either for internal use or for preventing invasions by outsiders. Communities sometimes request greater state intervention to improve forest condition or tenure security. It remains fairly uncommon, however, for states to recognize and support effective local institutions and practices and to integrate statutory and customary systems effectively.

In addition to recognizing the land and forest resources that are managed, at least to some extent, by customary practices, community tenure reforms also involve creating or recognizing a governance institution that represents the community. The size and boundaries of the forestland ceded by the state to local communities may coincide with an existing institution, but often a new level of governance and the formation of a new structure is required. This new institution is likely to play a central role in the allocation of rights to and benefits from forests, and a legitimate and effective institution ready to assume a new domain of powers on behalf of the community or at the larger scale may not exist prior to the reform.

Hence authority relations and the scale of their existence constitute a site of struggle and conflict. For example, the state and the community may not recognize the same actor as the legitimate community representative, or the recognized actor may not be accountable to the community. The construction of legitimate and accountable authority is a critical challenge for reforms involving communal or collective rights and, even in the absence of overt conflict, may involve delicate negotiation between traditional and modern political institutions. A trusted facilitator who can bridge those two cultures can be useful in such negotiations.

Other organizations beyond the community scale offer additional opportunities, and challenges, for representation. Given the failure of many states to carry out tenure changes fully or facilitate access to benefits from forests, community networks and other forms of collective action can be definitive for defending and increasing community rights and for improving market engagement. In fact, such networks can spend considerable human and financial resources just to defend community forest rights against competing interests, such as logging companies, colonists, conservation organizations and sometimes the state itself. With regard to market engagement, the most successful network, in the cases studied, is a producer federation that was set up specifically for this task; it appears much more difficult for political organizations to take on this additional and different set of challenges.

As political organizations, however, community networks such as the Federation of Community Forest User Groups, Nepal, have proven to be integral to stopping bureaucratic encroachment and negotiating new terms of engagement between communities and the state, especially where the state limits the rights granted to communities through regulation. This is partly related to the issue of co-management, mentioned above, but also goes beyond that. One type of regulation involves the macro-scale classification and zoning of forests, especially in Africa and some parts of Asia, whereby certain, usually higher-quality, forests fall under one classification (for industrial concessions or conservation) and lower-quality forests under another. This 'first cut' of defining who has access to what kind of forest often precedes the formal tenure reform, which may then recognize community rights only to forests with lower classifications, as in Cameroon, or to forests of lower value more generally, as in Nepal. Other types of regulations limit access to certain resources or require communities to jump through bureaucratic hoops to obtain permits. Though some regulation is important to protect forests for the future, existing legislation commonly includes rules that cannot be enforced and buttresses unnecessary and sometimes corrupt bureaucracies.

Regulations can make certain markets off-limits to communities, either through specific prohibitions or rules for compliance that are costly or otherwise prohibitive. Not all communities want to engage with markets and some believe that markets only allow others to capture rents from community resources and products. Most regulations affecting market access are skewed in favour of large industry or traders. Nevertheless, markets can also present opportunities and communities often engage with them informally if formal participation is too difficult or bureaucratic. Through regulations, the state can play a central role in affecting whether markets are opportunities or a danger for communities.

One factor affecting the outcomes of different forms of market engagement is community capacity. A common alternative to subsistence models, particularly in Latin America, has been the preconceived community forestry enterprise model, designed on the operating premises of large-scale logging, often for international markets. Whereas subsistence models may lead to much smaller livelihood improvements, the enterprise model can overwhelm communities with the demand to create new institutions and rapidly assume responsibilities and capacities, and it tends to foster external dependence. An emerging challenge is how to build an array of more appropriate, organic models that address both conservation and livelihood needs and are sustainable over the long term. The variety of cases examined here suggest that there is substantial room for policy improvements that would both build community capacity and address structural market distortions, such as legal and regulatory barriers, patron–client relationships and asymmetric information. Much less attention is usually paid to the latter.

Market conditions are another aspect that can affect outcomes for communities. Tenure reforms that facilitate engagement in timber markets provide the largest livelihood improvements as measured by change in income, particularly through the enterprise models mentioned above. But most of the reforms resulted in some kind of livelihood benefit when this was measured more broadly to include intangible benefits, such as empowerment or an end to outside intervention (such as state-authorized logging concessions) and access to new forest products and income. Reforms do not always improve resource access, however, and may even decrease it, at least temporarily and/or for some actors or products. This is because new rights are often accompanied by new restrictions, rules and responsibilities, and some resource users, particularly poor and marginalized groups, may be left out.

Most importantly, however, livelihood benefits are limited because of what happens after new rights have been granted on paper. During the process of implementation new rights are challenged and obstructed, both by state bureaucrats and by other powerful interest groups. And even when rights to forests are implemented in practice, little may be done to facilitate the exercise of those rights, such as through building community capacity, an enabling regulatory framework and beneficial market engagement, as discussed above.

Like livelihood improvements, which should be understood in light of these accompanying measures, changes in forest condition must be analysed in context. This is because forest conditions, in general, reflect multiple factors, some of which are outside the control of communities, such as pressure from loggers, miners, colonists or growing populations. It is notable, however, that forest conditions improved in cases where communities were given degraded lands and forests (particularly in Asia) and forest conditions did not decline under community management in several other cases, even when livelihoods improved.

Our research was also intended to explore implications of tenure reform for equity. The findings indicate that positive outcomes – avoiding elite capture, remedying gender and caste discrimination – were the result of specific policies and practices aimed at promoting equity, sometimes through positive discrimination. It is significant that the communities with the greatest apparent efforts to promote the rights of poor and disadvantaged groups are in Nepal, a country that has a powerful national movement and discourse promoting such policies. This alone does not remove structural disadvantages, but Nepal is clearly ahead of most of the other cases.

Central challenges

As the previous discussion has made clear, understanding reforms and their outcomes involves understanding three stages of the reform: the statutory change and its origin, the implementation of that change and the way in which the reform facilitated – or was combined with other factors to facilitate – improvements in livelihoods and forest condition. Each phase involves a different set of challenges and the statutory change is only the beginning of the reform process.

Statutory changes do not all promote sweeping changes in rights. The more ambitious reforms often emerged from grassroots demands – particularly for indigenous rights to traditional lands. In all cases, the implementation of reforms encounters delays and obstacles: competing interests and claims for the same forests or forest resources (whether from loggers, land grabbers, private industries or conservation organizations), lack of follow-through and the state's attempts to attenuate the rights granted. In fact, the state is charged with implementing statutory reforms, but another sector of the state may also be a competitor for resources. In particular, the cases studied demonstrate foot dragging in land titling, policy reversals, corruption and regulations of all kinds, as well as the failure of the state to defend new community rights from competing interests and intrusions.

Organized communities – and, in particular, community networks and federations – are better placed to defend their rights against these challenges. What actually gets implemented, then, is a result of struggle and opportunity combined, as reforms advance when communities and their allies take advantage of political moments. But political opportunities may arise before effective and accountable local management institutions have had time to form, which puts the benefits of reform at the risk of elite capture and the promotion or continuation of other inequities. It is not clear how to reconcile these contradictory needs. Hence the third stage of the reform, the facilitation and realization of benefits, faces two additional challenges: on the one hand, devising policies and programmes to bolster new opportunities and, on the other, supporting the creation of effective internal governance institutions and accountability mechanisms for decision-making and benefit distribution.

In general, across the three stages of reform, the obstacles facing communities can be grouped into three types: political, technical and conceptual. Most of the obstacles discussed so far are political and refer to competition for rights, resources and benefits from forests. They involve actors who oppose or interfere with reforms because they believe they have something to lose if communities are empowered, or who take advantage of reforms for their own gain: loggers, mining or petroleum companies that want resource rights, conservationists pushing for exclusive protected areas, bureaucrats who hold on to power and line their pockets by controlling decisions and resources, community leaders or elites who seek a disproportionate share of benefits. These political challenges require organized political responses.

Nevertheless, not all interference or problems with failed implementation or follow-through are due to political competition and corruption. Technical obstacles refer to capacity issues. The failure of the state to demarcate territories accurately, fairly or in a timely fashion, for example, may reflect a problem of human resources, such as experience or skill, or of funding. For their part, communities may not have experience in organized, collective forest management. Most reforms are new and constitute a steep learning process for all involved. Technical weaknesses, however, can be confused with more intentional delays and can also serve as a smokescreen for political interests of powerful actors. In addition, forest and environmental agencies are often reluctant to cede or share their technical roles with communities. Overcoming these weaknesses requires political will to obtain the knowledge or undertake the training required to move the reform process forward.

Conceptual obstacles refer to the extent to which communities are seen as, and given the chance to be, good forest stewards. Conceptual obstacles may also serve as a smokescreen for political interests, but there are real, legitimate concerns about the future of forests if communities are given greater rights. At the same time, from a rights perspective, and taking into account historical and traditional rights and past abuses of traditional peoples, communities should be granted their legitimate rights and should not be subject to laws and regulations other than those that apply to the rest of the population.

Some rights issues may have long-term consequences for – and beyond – forests. What are the economic, social, cultural and scientific consequences of declining customary practices and traditional knowledge due to use restrictions and the superimposition of state regulations over local rules? We may not know until it is too late. Transhumant pastoralism in Nepal's high hills constitutes a way of life for ethnic groups such as the Sherpas, Bhote and Tamang as well as a lucrative profession. It contributes to the national economy through the supply of milk, meat, draught animals and woollen goods, international trade and the identification of the region's species. But herder populations are declining as they are being banned from grazing their animals in forest areas and forced into smaller regions (Banjade and Paudel, 2008).

What are the consequences for forests? Outside 'experts' often appear to mistake sustainable local practices for degradation and take strong stances against an idea – fire, shifting cultivation, ranching, herding – without fully understanding each practice, its context or its long-term role in shaping forest landscapes; these ideas then become self-perpetuating and inaccurate narratives of degradation (Fairhead and Leach, 1996, 1998; Kull, 2004; Dove, 1983). In Nepal, pastoralists improve protection against forest fires and have superb ethnobotanic skills, traditional knowledge that may now be lost. Past evidence suggests there has been coordinated pasture management as well, with seasonal restrictions, rotational grazing and well-defined and mutually agreed rights.

We have already discussed at length the extent to which regulation – understood as over-regulation – interferes with new tenure rights, as the state retains the right to make important decisions about resource management. How much and what kind of regulation is really needed, under what circumstances and why, and how much is too much? Rather than starting from the perspective of state regulation, however, we propose starting from communities: what are local needs and practices and what potential do they have for sustainable, grassroots forest management? Fundamentally, if greater local control and appropriation is behind the principle of better and more sustainable management – and if greater long-term security promotes a long-term interest in sustaining resources – then to what extent do over-regulation and the retention of management rights interfere with its potential?

Fitzpatrick (2005) argues that the design of tenure reforms should be based on an assessment of the sources of tenure insecurity affecting communities (see Chapter 4). According to Fitzpatrick, the more external the insecurity, the less the state should interfere in internal affairs and, rather, focus on defending the perimeter of the community's customary area; the more internal, the greater the role for the state in mediating decisions over access.

A similar argument could be made regarding tenure reform and the causes of deforestation (see Table 10.1). The more external the causes of deforestation, the more the reform should seek to strengthen the community's exclusion and internal rule-making rights, while providing appropriate forums for negotiation with poor, external users (see Mwangi and Dohrn, 2008); the more internal, the greater the role for the state.

Current forest conditions should guide decisions regarding the extent to which recovery or maintenance of forest conditions (or management for certain products) is the priority. Internal incentives for forest maintenance, such as livelihood contributions or cultural values, should be reinforced and external pressures controlled. This constitutes another critical variable.

Table 10.1 merits some important caveats. First, it assumes that tenure rights have been granted or recognized and address underlying problems of insecurity. Second, it refers only to proximate causes of deforestation. The state itself may be an underlying cause of degradation if it promotes contradictory policies or specific policies that encourage forest clearing. These policies should be addressed as well. Third, external degradation may be a cause of internal degradation (Ribot, personal communication), if local people overexploit their

Table 10.1 *Degree and type of state regulatory role based on causes of deforestation and forest 'dependence'*

Contribution of (standing) forest to livelihoods or cultural reproduction	*Causes of deforestation/degradation*	
	External (or none)	*Internal*
Strong	No state intervention in community: state protects borders	Moderate state role: state facilitates rule enforcement
Weak	Moderate state role: state protects borders and facilitates organization and incentives to increase livelihood contribution*	High state role: greater state regulation of forest use (but communities still have right to participate in decisions)

* if desired by the community
Source: Elaborated by the authors based on ideas from Fitzpatrick (2005)

own resources rather than have them 'stolen' by outsiders. Hence external degradation should be addressed first and in this light: state facilitation of internal rule enforcement may not be needed.

At times, a strong role for the state will be justified, including through restrictions and regulations. But reforms should not be a way for the state to gain control over communities: forest departments still tend to blame local populations for degradation, failing to see communities as allies. Of particular concern are responsibilities that significantly constrain livelihoods, especially those of the poorest members of society; the failure to address or even recognize preexisting practices or the costs to communities of newly assigned responsibilities; corruption and rules that are unenforceable. The tenure reform should aim to reinforce or alter the incentive structure in favour of the use and conservation of forest products. The state should seek to provide incentives and increase capacities for local forest management, building on the potential knowledge, energy and indigenous organizational structures that are currently ignored or marginalized – an opportunity that has not yet been grasped and needs to be harmonized with formal management systems.

Forest tenure and emerging global concerns

The research conducted here examines cases in which communities have been granted greater statutory rights to forests. It demonstrates the benefits of these reforms, as well as some risks, and the many obstacles they have faced in implementation. Though formal statutory rights are not always needed and may at times (depending on how they are implemented) undermine some customary rights or a certain population's customary rights, formal rights appear to be particularly important in the face of competing interests with multiple stakeholders; and they may be increasingly important for the future security of forest rights – particularly with regard to climate change.

Climate change adds several new dimensions to an already complex framework of rights and resources. Forests both contribute to climate change and are affected by it, and forest-based populations are vulnerable both to direct climate change effects (ecological change, changing weather patterns, extreme events) and to competing interests for those forests or lands as mitigation schemes (such as carbon markets and bio-fuels expansion) mature.

The role of forests in influencing and responding to climate change is not fully understood (Science, 2008). Nevertheless, it is estimated that forests contribute more than 17 per cent to anthropogenic carbon emissions (IPCC, 2007). Higher global temperatures are expected to cause longer dry seasons and increases in forest fires and fire intensity, as they already have in some areas; they have also caused disruptions in seasonal patterns, such as rainfall or bird migrations, which may no longer be reliable indicators for making local land-use decisions (Macchi et al, 2008). In Nepal, climate change is leading to rising temperature, glacial retreat and changes in water availability. Extreme weather events such as hurricanes are also expected to increase; Hurricane Felix interrupted our research in Nicaragua. Changes in weather patterns and forest ecosystems will also affect the availability and distribution of wildlife and forest products.

The ability of populations to respond and adapt to these kinds of challenges depends to a large degree on policies. An International Union for the Conservation of Nature (IUCN) report on climate change concludes,

> *[I]nstitutions and policy makers play a key role in empowering indigenous and traditional peoples by securing and enhancing their entitlement to resources including land, water, biodiversity as well as health care, technology, education, information and power in order to improve their capacity to adapt to climate change and decrease their social and biophysical vulnerability. Where institutions fail to secure these entitlements, the resilience of indigenous and traditional peoples may decrease and the threshold, beyond which a system may not be able to adapt to environmental change, may be exceeded.* (Macchi et al, 2008, p22)

Though not all of the forest-based peoples studied here are indigenous, indigenous peoples constitute a particularly well-organized population globally that has issued its own formal declarations on these issues. One of the most important of these is the explicit priority given to food security. The Anchorage Declaration issued from the Indigenous People's Global Summit on Climate Change in early 2009 states:

> *In order to provide the resources necessary for our collective survival in response to the climate crisis, we declare our communities, waters, air, forests, oceans, sea ice, traditional lands and territories to be 'Food Sovereignty Areas,' defined and directed*

> *by Indigenous Peoples according to customary laws, free from extractive industries, deforestation and chemical-based industrial food production systems (i.e. contaminants, agro-fuels, genetically modified organisms).* (Anchorage Declaration, 2009)

But without secure and enforced land and resource rights, indigenous priorities for food security and cultural reproduction are challenged even further by climate change. In addition to ongoing demands for land and forests by competing actors, mitigation proposals also threaten forest peoples, such as through the expansion of bio-fuels and the reducing emissions from deforestation and degradation (REDD) schemes. Bio-fuels have increased the demand for land and though in theory they should not expand into forests (thereby negating any potential positive greenhouse gas emissions effects), this has occurred in some areas: in Indonesia, for example, the expansion of oil palm plantations has led to violence and repression and the takeover of indigenous lands without due process (Seymour, 2008).

Indigenous peoples have also issued their own response to REDD schemes, demanding that all initiatives 'secure the recognition and implementation of the human rights of Indigenous Peoples, including security of land tenure, ownership, recognition of land title according to traditional ways, uses and customary laws and the multiple benefits of forests for climate, ecosystems, and Peoples before taking any action' (Anchorage Declaration, 2009). As currently conceived, REDD strategies contemplate providing payments for avoided emissions from forest clearing and degradation (see Angelsen, 2008). REDD is a climate change strategy, however, not a poverty alleviation strategy, and the needs of poor people living in forests have not, at least not yet, been taken into account (Griffiths, 2008). Many people fear the consequences for local people and believe that REDD will not succeed without the support of indigenous groups (Brown et al, 2008; Griffiths, 2008; Macchi et al, 2008; Cotula and Mayers, 2009).

The problems are numerous. REDD and carbon markets introduce another layer of tenure rights to five pools of carbon – underground biomass, above-ground biomass, deadwood, litter and soil organic carbon – over the existing web of rights to land and forests. The question of who retains ownership over which carbon pool is significant in terms of the distribution of benefits from carbon marketing. In many cases the state might retain ownership. If, due to the actions of the local community, there are fewer forest fires and less deforestation, more carbon is retained in the biosphere; and if carbon stock increases, such as through the protection of natural regeneration, more carbon is captured from the atmosphere. But without clear rights over forests and carbon, it is likely that communities would not be able to claim benefits from REDD schemes, to the detriment of efforts to mitigate climate change.

Proposed REDD strategies fail even to acknowledge or address existing forest governance problems including, but not limited to, tenure as well as international human rights standards (Griffiths, 2008; Seymour, 2008). They are aimed at providing payments for avoided deforestation and hence could 'reward polluters with a history of forest destruction' but not those forest populations who already maintain and protect forest resources (Griffiths, 2008, p2). While this makes sense purely from an efficiency standpoint, it could undermine the legitimacy of the entire effort, foster conflict and provide perverse incentives for deforestation. Also, without secure tenure rights, local communities are 'vulnerable to dispossession – which could be a major concern if REDD increases land values and outside interest' (Cotula and Mayers, 2009, p3). Indigenous groups have demanded participation not only at the sub-national scale but also in global REDD negotiations.

The research presented throughout this book demonstrates that competition for forests and forestland is already fierce and that forest-based communities are often marginalized both in decision-making spheres and from access to forest resources and benefits. Even when they win new rights, serious challenges remain: for the implementation of rights in practice, for the defence of those rights and for the construction of the institutions necessary to exercise the rights, improve livelihoods and distribute benefits equitably. The state has dragged its heels on implementation of reforms, failed to defend community exclusion rights and retained decision-making powers over resource use. What do REDD schemes bring to this difficult scenario? If such schemes would prioritize protective strategies and severely restrict forest use, they would once again interfere with livelihood needs and impose formal restrictions and regulations over local rules and customs. If state officials have competed in the past with communities for resources as well as for decision-making power (and corruption continues to be a serious concern) this does not bode well for grassroots participation in, and the democratization of, strategies that require strict technical monitoring and compliance requirements and 'high levels of central coordination' (Cotula and Mayers, 2009, p2). [1]

The research also demonstrates the importance of follow-through in reforms and of a specific commitment to issues such as poverty, equity and representation. Substantial income benefits reached only those communities that had built the necessary institutions and market relations; gender and other equity issues had to be explicitly incorporated in reforms. New rights and benefits for collectives require attention to representation and authority relations; without serious attention to accountability, local 'authorities' may in fact be tools of the state or fail to distribute benefits.

Hence, secure tenure rights[2] are a necessary but not sufficient condition for protecting local populations and increasing resilience to threats from both climate change and mitigation efforts. They are also needed for these communities to actually benefit from REDD. At the same time, it is likely that insecure tenure contributes to climate change in at least two ways: by facilitating colonization and conversion of forests by 'outside' interests and by undermining traditional practices that have historically maintained forests (Anchorage Declaration, 2009). Secure tenure for groups living in forests, combined with exclusion rights protected by the state, could reduce colonization and conversion rates.

Given that the implementation of tenure rights in practice is still often tenuous, even when these rights are substantial, the land grab associated with bio-fuels plantations and possibly REDD schemes is likely to impede further – and possibly reverse past – progress in promoting community rights to forests. This reality cannot be ignored: the simple question of 'who owns the carbon?' provokes the issue. What strategies will competing interests use to undermine existing community rights? How will third parties try to take advantage of communities that have gained rights? What are the most effective strategies for communities to defend and deepen their rights, including participation in opportunities like REDD?

Indigenous groups and other forest-dependent populations must have a place at the bargaining table, both globally and nationally, to participate in the design, implementation and monitoring of climate change mitigation schemes. Within nations, the right to choose through free prior and informed consent (known as FPIC) should be required not only for indigenous peoples but all affected forest peoples. The importance of grassroots organization and higher-level networks cannot be overemphasized. Helping them, where needed, to understand the concepts, discourses, technicalities, biases and interests of climate change mitigation programmes and of their competitors, and providing the evidence from research to help sustain their arguments as they argue for their rights – this is the central role of their allies and supporters.

Future of reforms

We propose a tenure reform that starts with communities and builds on explicit agreements regarding rights and responsibilities as the basis of a workable system of forest governance. Ideally, resource decision-making will be located in the community and recognized as such, based on minimum standards for forest maintenance, and implemented with an emphasis on strengthening the collective governance structures in forest areas. Rights should be based on the recognition, but not the calcification, of customary rights and practices and the negotiation of conflict through transparent and accountable institutions. Zoning decisions, regarding different forest uses at scale, will be made with the understanding that high-quality forest areas should be designated for the recognition of community rights and include the informed participation of local rights holders. Alienation rights do not need to be granted to communities, but the state should not have the right to alienate these lands either, thus guaranteeing the permanence of rights and tenure security through strong tenure instruments.

The state will protect the rights of communities by guaranteeing their exclusion rights and upholding principles, such as free prior and informed consent, and will facilitate the negotiation mechanisms needed to address overlapping and seasonal resource rights of people external to communities. The state, together with other external actors, such as donors or NGOs, will facilitate the strengthening of local governance organizations and institutions for conflict resolution and the participation of communities in forest product markets. The 'models' of organization will be far-reaching and more akin to the nature and variety of community production patterns, allowing for the development of community-grown forest-based enterprises.

The state needs to review the organization and incoherence of its own policies across the sectors that affect forest tenure, management and governance. Ministries or agencies in agriculture, forestry, land reform, water, environment, minerals and hydrocarbons need to update their knowledge of the role of forests locally, nationally and globally, and rethink and reorganize their roles, policies and programmes. Since some deforestation and forest management problems stem from the state's own contradictory policies, the state agencies should reconcile and share their goals and support the capacity for local forest dwellers to become the protagonists of sustainable forest use and conservation.

Where continuing pressure on forestlands from colonists or internal conflict and lack of representation at the community level lead to deforestation, more emphasis is needed on understanding how current policies – subsidies for biofuel production, subsidies for industrial timber concessions, lack of instruments in forest planning to address the social realm of forest governance – may foster these problems. The combination of external interests and conflicting policies has often weakened and destroyed local governance without offering alternatives. Fostering and providing a central role for local decision-making in juggling and coordinating these often contradictory policies is a crucial step forward in governing forests. Promoting exposure between and discussion among sometimes antagonistic groups (colonists, indigenous, traditional forest peoples) seeking access to forestland could be more advantageous than pitting them against each other. In cases where interests in alternative land uses are desired, communities themselves need to be a part of the decision-making for compensation or alternative proposals to determine the real value of their assets.

Given past experience, we recognize that no such ideal states or policies exist; what happens in practice will instead be defined by social and political processes of negotiation and contestation. Hence what we are proposing is a road map for communities and community organizations and their advocates... for the future of forest tenure reform.

Notes

1. Central coordination is needed to guarantee 'strong and fair rules and institutions, macroeconomic and agricultural policies in tune with forest policies, effective monitoring' (Cotula and Mayers, 2009, p2).
2. We also recognize that secure tenure can lead to forest conversion for more profitable uses (Tacconi, 2007a), as expressed elsewhere in this book.

References

Acharya, R. P. (2005) 'Socio-economic impacts of community based forest enterprises in mid hills of Nepal–Case Study from Dolakha district', *Banko Janakari*, vol 15, no 2, pp43–47

ACOFOP (2005) *Guía básica para los habitantes de las comunidades forestales*, Petén, Guatemala

ACOFOP-CIFOR (2007) 'Informe para el proyecto CIFOR-ACOFOP: nuevas tendencias y procesos que influyen en el manejo comunitario forestal en la Zona de Usos Múltiples Reserva de Biósfera Maya en Petén, Guatemala', Iliana Monterroso, Guatemala

Adams, W. M. (2004) *Against Extinction: The Story of Conservation*. Earthscan, London

Adhikari, K. P. (2007) 'A review of literature on forest tenures and impact on livelihood, forest condition, income and equity (LIFE) in Asia', draft report, CIFOR, Bogor, Indonesia

Adhikari, B. and Lovett, J. C. (2006) 'Transaction costs and community-based natural resource management in Nepal', *Journal of Environmental Management*, vol 78, pp5–15

Adhikari, B., Falco, S. D. and Lovett, J. C. (2004) 'Household characteristics and forest dependency: evidence from common property forest management in Nepal', *Ecological Economics*, vol 48, pp245–257

Agarwal, B. (2001) 'Participatory exclusion, community forestry and gender: an analysis for South Asia and a conceptual framework', *World Development*, vol 29, no 10, pp1623–1648

Agbosu, L. K. (2000) 'Land law in Ghana: contradiction between Anglo-American and customary tenure conceptions and practices', Working Paper 33, Land Tenure Center, University of Wisconsin–Madison, USA

Agrawal, A. (2001) 'Common property institutions and sustainable governance of resources', *World Development*, vol 29, no10, pp1649–1672

Agrawal, A. and Gibson, C. (1999) 'Enchantment and disenchantment: the role of community in natural resource conservation', *World Development*, vol 27, no 4, pp629–649

Agrawal, A. and Ostrom, E. (2001) 'Collective action, property rights, and decentralization in resource use in India and Nepal', *Politics and Society*, vol 29, no 4, pp485–514

Agrawal, A. and Chhatre, A. (2006) 'Explaining success on the commons: community forest governance in the Indian Himalaya', *World Development*, vol 34, no 1, pp149–166

Albornoz, M. A. and Toro, M. (2008) 'Acceso a la tierra y manejo forestal en la economía extractivista del norte boliviano', unpublished draft, CEDLA and CIFOR, La Paz, Bolivia

Albornoz, M., Cronkleton, P. and Toro, M. (2008) *Estudio regional Guarayos: historia de la configuración de un territorio en conflicto*. CEDLA and CIFOR, Santa Cruz, Bolivia

Alden Wily, L. (2004) 'Can we really own the forest? A critical examination of tenure development in community forestry in Africa', paper prepared for 10th Biennial Conference, International Association for the Study of Common Property (IASCP), Oaxaca, Mexico, August 9–13

Alden Wily, L. (2008) 'Custom and commonage in Africa: rethinking the orthodoxies', *Land Use Policy*, vol 25, pp43–52

Alegret, R. (2003) 'Evolución y tendencias de las reformas agrarias en América Latina', in *Land Reform, Land Settlement and Cooperatives*. Food and Agriculture Organization, available at www.fao.org/docrep/006/J0415T/j0415t0b.htm#bm11 (last accessed September 2009)

Anaya, S. J. and Grossman, C. (2002) 'The case of Awas Tingni v. Nicaragua: a new step in the international law of indigenous peoples', *Arizona Journal of International and Comparative Law*, vol 19, no 1, pp1–15

Anchorage Declaration (2009) 'Indigenous Peoples' Global Summit on Climate Change, consensus agreement', Anchorage Alaska, April 24, available at www.indigenoussummit.com/servlet/content/declaration.html (last accessed May 2009)

Angelsen, A. (1995) 'Shifting cultivation and "deforestation": a study from Indonesia', *World Development*, vol 23, no 10, pp1713–1729

Angelsen, A. (ed.) (2008) *Moving ahead with REDD: issues, options and implications*, CIFOR, Bogor, Indonesia

Angelsen, A. and Wunder, S. (2003) 'Exploring the forest-poverty link: concepts, issues and research implications', Occasional Paper 40, CIFOR, Bogor, Indonesia

Ankersen, T. and Ruppert, T. (2006) 'Tierra y Libertad, the social function doctrine and land reform in Latin America', *Tulane Environmental Law Journal*, vol 19, pp69–120

Antinori, C. (2005) 'Vertical integration in the community forestry enterprises of Oaxaca', in D. Bray, L. Merino-Pérez and D. Barry (eds) *The Community Forestry of Mexico: Managing for Sustainable Landscapes*. University of Texas Press, Austin, Texas, USA

Aramayo Caballero, J. (2004) *La Reconstitución del Sistema Barraquero en el Norte Amazonico: Analisis jurídico del Decreto Supremo No 27572*, CEJIS, Santa Cruz, Bolivia

Argüello, A. (2008) 'Cadena de valor de la madera de la cooperativa Kiwatigni en Layasiksa-RAAN' (The forest value chain of the Kiwatingni Cooperative in Layasiksa, RAAN), unpublished report, CIFOR and Masangni, Managua

Armitage, D. R., Plummer, R., Berkes, F., Arthur, R., Charles, A., Davidson-Hunt, I., Diduck, A., Doubleday, N., Johnson, D., Marschke, M., McConney, P., Pinkerton, E. and Wallenberg, E. (2009) 'Adaptive comanagement for social-ecological complexity', *Frontiers and Ecology and Environment*, vol 7, no 2, pp95–102

Ash-Garner, R. and Zald, M. (1987) 'The political economy of social movements', in M. Zald and J. D. McCarthy (eds) *Social Movements in an Organizational Society*. Transaction Books, New Brunswick, New Jersey, USA

Assies, W. (2008) 'From rubber estate to simple commodity production: agrarian struggles in the northern Bolivian Amazon', *The Journal of Peasant Studies*, vol 29, no 3–4, pp83–130

Ayine, D. (2008) 'Social responsibility agreements in Ghana's forestry sector', Developing legal tools for citizen empowerment series, IIED, London

Bae, M. H. M. (2005) *Global Patterns of Alienation and Devolution of Indigenous and Tribal Peoples' Land*. World Bank, Washington, DC

Baird, I. G. and Shoemaker, B. (2005) *Aiding or Abetting? Internal Resettlement and International Aid Agencies in the Lao PDR*. Probe International, Toronto, Ontario, Canada

Baland, J. M. and Platteau, J. P. (1996) *Halting Degradation of Natural Resources: Is There a Role for Rural Communities?* FAO and Clarendon Press, Oxford

Ballard, P., Habib, A., Valodia, I. and Zuern, E. (2003) *Globalization, Marginalization, and Contemporary Social Movements in South Africa*. Centre for Civil Society, University of KwaZulu Natal, Durban

Bampton, J. and Cammaert, B. (2007) 'How can timber rents better contribute to poverty reduction through community forestry in the Terai region of Nepal?', *Journal of Forest and Livelihood*, vol 6, no 1, pp28–47

Banjade, M. R. and Paudel, N. S. (2008) *Suspa community forest users group, Dokakha*. CIFOR and Forest Action, Kathmandu

Barr, C., Wollenberg, E., Limberg, G., Anau, N., Iwan, R., Sudana, I. M, Moeliono, M. and Djogo, T. (2001) *The Impacts of Decentralisations on Forests and Forest-Dependent Communities in Malinau District, East Kalimantan*. CIFOR, Bogor, Indonesia

Barr, C., Brown, D., Casson, A. and Kaimowitz, D. (2002) 'Corporate debt and the Indonesian forestry sector', in C. J. P. Colfer and I. A. P. Resosudarmo (eds) *Which Way Forward? People, Forests and Policymaking in Indonesia*. CIFOR and Resources for the Future, Washington, DC

Barr, C., Resosudarmo, I. A. P., Dermawan, A. and McCarthy, J. (eds) (2006) *Decentralization of Forest Administration in Indonesia: Implications for Sustainability, Economic Development and Community Livelihoods*. CIFOR, Bogor, Indonesia

Barry, D. and Meinzen-Dick, R. (2008) 'The invisible map: community tenure rights', paper presented at Conference of the International Association for the Study of Commons (IASC), Cheltenham, UK

Barry, D. and Monterroso, I. (2008) 'Institutional change and community forestry in the Mayan Biosphere Reserve Guatemala', paper presented at Conference of the International Association for the Study of Commons (IASC), Cheltenham, UK

Barry, D. and Taylor, P. (2008) *An Ear to the Ground: Tenure Changes and Challenges for Forest Communities in Latin America*. CIFOR and Rights and Resources Initiative, Washington, DC, available at www.rightsandresources.org/publication_details.php?publicationID=929 (last accessed September 2009)

Becker, L. (2001) 'Seeing green in Mali's woods: colonial legacy, forest use, and local control', *Annals of the Association of American Geographers*, vol 91, no 3, pp504–26

Belcher, B., Ruíz-Pérez, M. and Achdiawan, R. (2005) 'Global patterns and trends in the use and management of commercial NTFPs: implications for livelihoods and conservation', *World Development*, vol 33, no 9, pp1435–1452

Benjamin, C. (2008) 'Legal pluralism and decentralization: natural resource management in Mali', *World Development*, vol 36, no 11, pp2255–2276

Bennett, C. P. A. (2002) 'Responsibility, accountability, and national unity in village governance', in C. J. P. Colfer and I. A. P. Resosudarmo (eds) *Which Way Forward? People, Forests and Policymaking in Indonesia*. CIFOR and Resources for the Future, Washington, DC

Berkes, F., George, P. and Preston, R. (1991) 'Comanagement: the evolution of the theory and practice of joint administration of living resources', *Alternatives*, vol 18, no 2, pp12–18

Berry, S. (1993) *No Condition is Permanent: The Social Dynamics of Agrarian Change in Sub-Sahara Africa*. University of Wisconsin Press, Madison, Wisconsin, USA

Berry, S. S. (2001) *Chiefs Know Their Boundaries: Essays on Property, Power and the Past in Asante, 1896–1996*. David Philip, Cape Town

Bhattarai, B. (2006) 'Widening the gap between terai and hill farmers in Nepal: the implications of the New Forest Policy 2000', in S. Mahanty, J. Fox, M. Nurse, P. Stephen and L. McLees (eds) *Hanging in the Balance: Equity in Community-Based Natural Resource Management in Asia*. RECOFTC, Bangkok, Thailand, and East-West Center, Honolulu, Hawaii, USA

Blaikie, P. (1985) *The Political Economy of Soil Erosion in Developing Countries*. Longman, London

Bojanic, A. (2001) *Balance is Beautiful: Assessing Sustainable Development in the Rain Forest of the Bolivian Amazon*. University of Utrecht, PROMAB, The Netherlands

BOLFOR II (2007) 'Beneficios del Plan de Manejo Forestal de la Comunidad Cururú, Gestion 2006', Proyecto BOLFOR II, Santa Cruz, Bolivia

Boni, S. (2005) *Clearing the Ghanaian Forest: Theories and Practices of Acquisition Transfer and Utilisation of Farming Titles in the Sefwi-Akan Area*. Institute of African Studies, Legon, Accra

Bray, D. B. and Anderson, A. B. (2006) 'Global conservation non-governmental organizations and local communities: perspectives on programs and project implementation in Latin America', Working Paper 1, Conservation and Development Series, Latin American and Caribbean Center, Florida State University, Tallahassee, Florida, USA

Bray, D. B., Merino-Pérez, L. and Barry, B. (eds) (2005) *The Community Forests of Mexico, Managing for Sustainable Landscapes*. University of Texas Press, Austin, Texas, USA

Bray, D. B., Antinori, C. and Torres-Rojo, J. M. (2006) 'The Mexican model of community forest management: the role of agrarian policy, forest policy and entrepreneurial organization', *Forest Policy and Economics*, vol 8, pp470–484

Bray, D., Duran, B., Ramos, V., Mas, J., Velázquez, A., McNab, R., Barry, D. and Radachowsky, J. (2008) 'Tropical deforestation, community forests and protected areas in the Maya forest', *Ecology and Society*, vol 13, no 2, available at www.ecologyandsociety.org/vol13/iss2/art56/ (last accessed April 2009)

Brockington, D. and Igoe, J. (2006) 'Eviction for conservation: a global overview', *Conservation and Society*, vol 4, no 3, pp424–470

Brockington, D., Igoe, J. and Schmidt-Soltau, K. (2006) 'Conservation, human rights and poverty reduction', *Conservation Biology*, vol 20, pp250–252

Bromley, D. W. (2004) 'Reconsidering environmental policy prescriptive consequentialism and volitional pragmatism', *Environmental and Resource Economics*, vol 28, no 1, pp73–99

Brown, D., Seymour, S. and Peskett, L. (2008) 'How do we achieve REDD co-benefits and avoid doing harm?' in A. Angelsen (ed.) *Moving Ahead with REDD: Issues, Options and Implications*. CIFOR, Bogor, Indonesia

Bruce, J. (1998) 'Review of tenure terminology', Tenure Brief, Land Tenure Center, University of Wisconsin, Madison, Wisconsin, USA

Campese, J., Sunderland, T., Greiber, R. and Oviedo, G. (2009) *Rights Based Approaches: Exploring Issues and Opportunities for Conservation*, CIFOR and IUCN, Bogor, Indonesia

Carlsson, L. and Berkes, F. (2005) 'Comanagement: concepts and methodological implications', *Journal of Environmental Management*, vol 75, pp65–76

Carvalheiro, K. (2008) 'Análise da Legislação para o Manejo Florestal por Pequenos Produtores na Amazônia Brasileira' (Analysis of forest management legislation for small producers in the Brazilian Amazon), unpublished draft, CIFOR, Belem, Pará, Brazil

CCARC (Caribbean and Central American Research Council) (2000) 'Diagnóstico general sobre la tenencia de la tierra en las comunidades indígenas de la Costa Atlántica', reprinted in A. Rivas and R. Broegaard (eds) (2006) *Demarcación Territorial de la Propiedad Comunal en la Costa Caribe de Nicaragua*. MultiGrafic, Managua

Cernea, M. (1997) 'The risks and reconstruction model for resettling displaced populations', *World Development*, vol 25, no 10, pp1569–1587

Cernea, M. (2006) 'Population displacement inside protected areas: a redefinition of concepts in conservation policies', *Policy Matters*, vol 14, pp8–26

Chapagain, D. P., Kanel, K. R. and Regmi, D. C. (1999) 'Current policy and legal context of the forestry sector with reference to the community forestry programme in Nepal', working paper submitted to Nepal-UK Community Forestry Project, Kathmandu, Nepal

Chape, S., Blyth, S., Fish, L., Fox, P. and Spalding, M. (2003) 2003 *United Nations List of Protected Areas*, IUNC, Gland and UNEP World Conservation Monitoring Centre, Cambridge

Chapin, M., Lamb, A. and Threlkeld, B. (2005) 'Mapping indigenous lands', *Annual Review of Anthropology*, vol 34, pp619–638

CIFOR (1999) *Criteria & Indicators Toolbox*. 9 vols, CIFOR, Bogor, Indonesia

Clay, J. W. (1988) 'Indigenous peoples and tropical forests: models of land use and management from Latin America', Report 27, Cultural Survival Inc., Cambridge, Massachusetts, USA

CNS (Conselho Nacional dos Seringueiros) (2005) 'Populações Extrativistas da Amazônia: processo histórico, conquistas sócio-ambientais e estratégia de desenvolvimento econômico', Conselho Nacional dos Seringueiros, Belém, Brasil

Cohen, J. L. (1983) 'Rethinking social movements', *Berkeley Journal of Sociology*, vol 28, pp97–114

COICA (Coordinator of Indigenous Organizations of the Amazon Basin) (2003) 'La visión de la organización indígena COICA sobre áreas protegidas', WRM, Movimiento Mundial por los Bosques Tropicales, available at www.wrmo.org.uy/73/COICA.htm (last accessed June 2009)

Colchester, M. (2000a) 'Self-determination or environmental determinism for indigenous peoples in tropical forest conservation', *Conservation Biology*, vol 14, no 5, pp1365–1367

Colchester, M. (2000b) *Indigenous Peoples and the New 'Global Vision' on Forests: Implications and Prospects*. Forest and People Program, London

Colchester, M. (2004) 'Conservation policy and indigenous peoples', *Environmental Science and Policy*, vol 7, pp145–153

Colchester, M., Jiwan, N., Andiko, Sirait, M., Firdaus, A. Y., Surambo, A. and Pane, H. (2006) *Promised Land: Palm Oil and Land Acquisition in Indonesia: Implications for Local Communities and Indigenous People*, Moreton-in-Marsh, England, and Bogor, Indonesia: Forest Peoples Programme and Perkumpulan Sawit Watch

Colfer, C. J. P. (1985a) 'Female status and action in two Dayak communities', in M. Goodman (ed.) *Women in Asia and the Pacific: Toward an East-West Dialogue*. University of Hawaii Press, Honolulu, Hawaii, USA

Colfer, C. J. P. (1985b) 'On circular migration: from the distaff side', in G. Standing (ed.) *Labour Circulation and the Labour Process*. Croom Helm Ltd, London

Colfer, C. J. P. (1991) *Toward Sustainable Agriculture in the Humid Tropics: Building on the Tropsoils Experience in Indonesia*, Tropsoils Technical Bulletin No. 91/02, Raleigh, North Carolina, USA

Colfer, C. J. P. (ed) (2005) *The Equitable Forest: Diversity, Community and Resource Management*. CIFOR and Resources for the Future, Washington, DC

Colfer, C. J. P. with Dudley, R.G. (1993) 'Shifting cultivators of Indonesia: managers or marauders of the forest? Rice production and forest use among the Uma' Jalan of East Kalimantan', Community Forestry Case Study Series 6, Food and Agriculture Organization of the United Nations, Rome

Colfer, C. J. P. and Byron, Y. (2001) *People Managing Forests: The Links Between Human Well-Being and Sustainability*. CIFOR and Resources for the Future, Washington, DC

Colfer, C. J. P. and Capistrano, D. (eds) (2005) *The Politics of Decentralization: Forests, Power and People*. Earthscan, London

Colfer, C. J. P., Gill, D. and Agus, F. (1988) 'An indigenous agroforestry model from West Sumatra: a source of insight for scientists', *Agricultural Systems*, 26, pp191–209

Colfer, C. J. P., Newton, B. and Herman (1989) 'Ethnicity: an important consideration in Indonesian agriculture', *Agriculture and Human Values,* VI (no 3). Earlier version published in *Proceedings Centre for Soils Research Technical Meetings*, Bogor, Indonesia (1986)

Colfer, C. J. P., Peluso, N. L. and Chin, S. C. (1997) 'Beyond slash and burn: building on indigenous management of Borneo's tropical rain forests', *Advances in Economic Botany*, vol 11, New York Botanical Garden, Bronx, New York, USA

Colfer, C. J. P., Dahal, G. R. and Capistrano, D. (2008a) *Lessons from Forest Decentralisation in Asia Pacific: Money, Justice and Quest for Good Governance*. Earthscan Publications, London

Colfer, C. J. P., Dudley, R. G. and Gardner, R. (2008b) *Forest Women, Health and Childbearing. Human Health and Forests: A Global, Interdisciplinary Overview*. Earthscan, London

Contreras-Hermosilla, A. (2001) 'Forest law compliance: an overview', World Bank, Washington, DC

Conyers, D. (1983) 'Decentralization: the latest fashion in development administration?' *Public Administration and Development*, vol 3, pp97–109

Cotula, L. and Mayers, M. (2009) 'Tenure in REDD: start-point or afterthought?' International Institute for Environment and Development, London

Cousins, B. (2007a) 'Agrarian reform and the "two economies": transforming South Africa's countryside', in L. Ntsebeza and R. Hall (eds) *The Land Question in South Africa*, HSRC Press, Cape Town

Cousins, B. (2007b) 'More than socially embedded: the distinctive character of "Communal Tenure" regimes in South Africa and its implications for land policy', *Journal of Agrarian Change*, vol 7, no 3, pp281–315

Cousins, B. and Sjaastad, E. (eds) (2008) Land Use Policy, Special Issue, vol 26, no 1

CRAAN (2007) 'Ayuda memoria: Asamblea territorial de Tasba Raya, Waspam, Llanos y Río Abajo y el Territorio MISRAT, Municipio de Waspam Río Coco', Consejo de la Región Autónoma Atlántico Norte, Bilwi, Nicaragua, May 19

Cramb, R. A., Pierce Colfer, C. J., Dressler, W., Laungaramsri, P., Trung Le, Q., Mulyoutami, E., Peluso, N. L. and Wadley, R. L. (2009) 'Swidden transformations and rural livelihoods in Southeast Asia', *Human Ecology* 37, pp323–346

Cronkleton, P. and Pacheco, P. (2008a) 'Changing policy trends in the emergence of Bolivia's Brazil nut sector', in S. Laird, R. McLain and R. Wynberg (eds) *Non-Timber Forest Products Policy: Frameworks for the Management, Trade and Use of NTFPs*. Earthscan, London

Cronkleton, P. and Pacheco, P. (2008b) 'Communal tenure policy and the struggle for forest land in the Bolivian Amazon', paper presented at Conference of the International Association for the Study of Commons (IASC), Cheltenham, UK

Cronkleton, P., Taylor, P., Barry, D., Stone-Jovicich, S. and Schmink, M. (2008) 'Environmental governance and the emergence of forest-based social movements', Occasional Paper 49, CIFOR, Bogor, Indonesia

Cronkleton, P., Pacheco, P., Ibarguen, R. and Albornoz, M. (2009) *Reformas en la tenencia de la tierra y los bosques: La gestión comunal en las tierras bajas de Bolivia*. CIFOR and CEDLA, La Paz, Bolivia

Curran, B., Sunderland, T., Maisels, F., Oates, J., Asaha, S., Balinga, M., Defo, L., Dunn, A., Telfer, P., Usongo, L., von Loebenstein, K. and Roth, P. (in press) 'Are Central Africa's protected areas displacing hundreds of thousands of rural poor?' *Conservation and Society*

Da Rocha, B. J. and Lodoh, C. H. K. (1999) 'Ghana land law and conveyancing', Anansesem Publications, Ghana

Dahal, G. R. and Adhikari, K. P. (2008) 'Bridging, linking and bonding social capital in collective action', Working Paper 79, CAPRi, Washington, DC

Dahal, G. R. and Chapagain, A. (2008) 'Community forestry in Nepal: decentralized forest governance', in C. J. P. Colfer, G. R. Dahal and D. Capistrano (eds) *Lessons from Forest Decentralization: Money, Justice and the Quest for Good Governance*. CIFOR and Earthscan, London

Dana, S. T. and Fairfax, S. K. (1980) *Forest and Range Policy: Its Development in the United States*. McGraw-Hill, New York

Davis, S. H. and Wali, A. (1994) 'Indigenous land tenure and tropical forest management in Latin America', *Ambio*, vol 23, no 8, pp485–490

de Camino, R. (2000) 'Algunas consideraciones sobre el manejo forestal comunitario y su situación en América Latina, Taller: Manejo Forestal Comunitario y Certificación en América Latina – estado de experiencias actuales y perspectivas futuras', Santa Cruz, Bolivia 21–27 enero

de Janvry, A. (1981) *The Agrarian Question and Reformism in Latin America*. Johns Hopkins University Press, Baltimore, Maryland, USA

de Jong, W., Ruiz, S. and Becker, M. (2006) 'Conflicts and communal forest management in northern Bolivia', *Forest Policy and Economics*, vol 8, pp447–457

Deininger, K. and Binswanger, H. (2001) 'The evolution of the World Bank's policy', in A. de Janvry, G. Gordillo, J. Platteau and E. Sadoulet (eds) *Access to Land, Rural Poverty, and Public Action*. Oxford University Press, Oxford and New York

Diaw, C. (1997) 'Si, Nda Bot, and Ayong: shifting cultivation, land use, and property rights in southern Cameroon', vol. 21e, Rural Development Forestry Network Paper, ODI, London

Diaw, M. C. (2005) 'Modern economic theory and the challenge of embedded tenure institutions: African attempts to reform local forest policies', in S. Kant and R. A. Berry (eds) *Sustainability, Institutions and Natural Resources: Institutions for Sustainable Forest Management*. Springer, The Netherlands

Diaw, M. C. (2009) 'Elusive meanings: decentralization, conservation and local democracy', in L. German, A. Karsenty and A.-M. Tiani (eds) *Governing Africa's Forests in a Globalized World*. CIFOR and Earthscan, London

Diaw, M. C., Aseh, T. and Prabhu, R. (eds) (2008) *In Search of Common Ground: Adaptive Collaborative Management in Cameroon*. CIFOR, Bogor, Indonesia

Dixon, J. A. and Sherman, P. B. (1991) *Economics of Protected Areas: A New Look at Costs and Benefits*. Earthscan, London

Dizon, J. T., Pulhin, J. M. and Cruz, R. V. O. (2008) 'Improving equity and livelihoods in community forestry: the case of the Kalahan Educational Foundation in Imugan, Sta. Fe, Nueva Vizcaya, Philippines', project report, CIFOR and RRI, Bogor, Indonesia

Donovan, J., Stoian, D., Macqueen, D. and Grouwels, S. (2006) 'The business side of sustainable forest management: development of small and medium forest enterprises for poverty reduction', *Natural Resource Perspectives*, vol 104

Donovan, J., Stoian, D., Grouwles, S., Macqueen, D., van Leeuwen, A., Boetekees, G. and Nicholson, K. (2008a) 'Towards an enabling environment for small and medium forest enterprise development', CATIE, FAO, IIED, SNV, ICCO, San José, Costa Rica

Donovan, J., Stoian, D. and Poole, N. (2008b) 'Global review of rural community enterprises: the long and winding road for creating viable businesses and potential shortcuts', CATIE, SOAS, San José, Costa Rica

Dove, M. R. (1983) 'Theories of swidden agriculture, and the political economy of ignorance', *Agroforestry Systems*, vol 1, no 2, pp85–99

Dowie, M. (2005) 'Conservation refugees: when protecting nature means kicking people out', *Orion Magazine*, November/December, available at www.orionmagazine.org/index.php/articles/article/161 (last accessed September 2009)

Dugan, P. and Pulhin, J. (2006) 'Forest harvesting in community-based forest management (CBFM) in the Philippines: simple tools versus complex procedures', in R. Oberndorf, P. Durst, S. Mahanty, K. Burslem and R. Suzuki (eds) 'A Cut for the Poor', Proceedings of International Conference on Managing Forests for Poverty Reduction: Capturing Opportunities in Forest Harvesting and Wood Processing for the Benefit of the Poor, Ho Chi Minh City, Vietnam, 3–6 October. FAO RAP Publication 2007/09 and RECOFTC Report No. 19, FAO and RECOFTC, Bangkok

Eckersley, R. (1992) *Environmentalism and Political Theory: Toward an Ecocentric Approach*. University College London Press, London

Edmunds, D. and Wollenberg, E. (eds) (2003) *Local Forest Management*. Earthscan, London

Edmunds, D., Wollenberg, E., Contreras, A., Dachang, L., Kelkar, G., Nathan, D., Sarin, M. and Singh, N. (2003) 'Introduction', in D. Edmunds and E. Wollenberg (eds) *Local Forest Management: The Impacts of Devolution Policies*. Earthscan, London

El-Ghonemy , M. R. (2003) 'Land reform development challenges of 1963–2003 continue into the twenty-first century in land reform, land settlement and cooperatives', Economic and Social Development Department, Food and Agriculture Organization, Rome, available at www.fao.org/DOCREP/006/J0415T/j0415t05.htm#bm05 (last accessed September 2009)

Elías, S. and Wittman, H. (2005) 'State, forest and community: decentralization of forest administration in Guatemala', in C. J. P. Colfer and D. Capistrano (eds) *The Politics of Decentralization: Forests, Power and People*. Earthscan, London

Elías, S., Larson, A. and Mendoza, J. (2009) *Tenencia de la tierra, bosques y medios de vida en el Altiplano Occidental de Guatemala*. CIFOR and FAUSAC, Guatemala

Elliott, C. (1996) 'Paradigms of forest conservation', *Unasylva*, vol 47, p187

Ellsworth, L. (2002) *A Place in the World: Tenure Security and Community Livelihoods, A Literature Review,* Forest Trends, Washington, DC, and Ford Foundation, New York

Enters, T., Qiang, M. and Leslie, R. N. (2003) *An Overview of Forest Policies in Asia.* Food and Agriculture Organization, Bangkok

Fairhead, J. and Leach, M. (1996) *Misreading the African Landscape: Society and Ecology in a Forest-Savanna Mosaic.* Cambridge University Press, Cambridge and New York

Fairhead, J. and Leach, M. (1998) *Reframing Deforestation: Global Analyses and Local Realities – Studies in West Africa.* Routledge, London

FAO (2005) *Global Forest Resources Assessment.* Food and Agriculture Organization, Rome

FAOSTAT (2007) FAO Statistical Database, available at http://faostat.fao.org/site/535/DesktopDefault.aspx?PageID=535 (last accessed September 2009)

Faulks, D. (1999) *Political Sociology, A Critical Introduction.* Edinburgh University Press, Edinburgh, Scotland

Fay, D. (2008) '"Traditional authorities" and authority over land in South Africa', paper presented at Conference of the International Association for the Study of the Commons (IASC), July 14–18, Cheltenham, England

Fay, C. and Michon, G. (2003) 'The contribution of plantation and agroforestry to rural livelihoods: redressing forestry hegemony – where a forestry regulatory framework is best replaced by an agrarian one', paper presented at International Conference on Rural Livelihoods, Forests and Biodiversity, May 19–23, Bonn, Germany

FECOFUN (1999) 'Samayik Prakashan: Kathmandu', Network of Community Forestry Users, Nepal (FECOFUN), vol 3, Kathmandu

FECOFUN (2002) 'Report of Third National Council Meeting', FECOFUN, Kathmandu

Feeny, D., Berkes, F., McCay, B. and Acheson, J. (1990) 'The tragedy of the commons: twenty-two years later', *Human Ecology*, vol 18, no 1, pp1–19

Fernow, B. (1911) *The History of Forestry.* 3rd ed., University of Toronto Press, Toronto, Ontario, Canada

Ferroukhi, L. (ed.) (2004) *Municipal Forest Management in Latin America.* CIFOR and IDRC, Bogor, Indonesia

FES (2007) 'Annual Report 2006–2007', Foundation for Ecological Security, Anand, Gujarat, India

Fischer, R. (1995) *Collaborative Management of Forests for Conservation and Development.* IUCN and WWF, Gland, Switzerland

Fisher, R. J., Maginnis, S., Jackson, W. J., Barrow, E. and Jeanrenaud, S. (2005) *Poverty and Conservation: Landscapes, People and Power.* World Conservation Union, IUCN, Zurich, Switzerland

Fitriana, J. R. (2008) 'Landscape and farming system in transition: case study in Viengkham District, Luang Prabang Province, Lao PDR', Agronomy and Agro-Food Program, Institut des Régions Chaudes-Supagro, Montpelier, France

Fitzpatrick, D. (2005) '"Best practice" options for the legal recognition of customary tenure', *Development and Change*, vol 36, no 3, pp449–475

Fitzpatrick, D. (2006) 'Evolution and chaos in property rights systems: the third world tragedy of contested access', *Yale Law Journal*, vol 115, pp996–1048

Forsyth, T. (2007) 'Are environmental social movements socially exclusive? An historical study from Thailand', *World Development*, vol 35, no 12, pp2110–2130

Fortmann, L. (1987) 'Tree tenure: an analytical framework for agroforestry projects', ICRAF, Nairobi

FSI (2003) *The State of Forest Report*. Forest Survey of India, Dehradun

Fulcher, M. B. (1982) 'Dayak and transmigration in East Kalimantan', *Borneo Research Bulletin*, vol 14, no 1, pp14–23

Geist, H. J. and Lambin, F. (2002) 'Proximate causes and underlying driving forces of tropical deforestation', *Bioscience*, vol 52, no 2, pp143–150

Gentle, P., Acharya, K. P. and Dahal, G. R. (2007) 'Advocacy campaign to improve governance in community forestry: a case from western Nepal', *Journal of Forest and Livelihoods*, vol 6, no 1, pp59–69

German, L., Karsenty, A. and Tiani, A.M. (eds) (2009) *Governing Africa's Forests in a Globalized World*. CIFOR and Earthscan, London

Ghate, R. and Beasley, K. (2007) 'Aversion to relocation: a myth?', *Conservation and Society*, vol 5, no 3, pp331–334

Gibson, C., McKean, M. and Ostrom, E. (eds) (2000) *People and Forests: Communities, Institutions, and Governance*. MIT Press, Cambridge, Massachusetts, USA

Gilmour, D., Malla, Y. and Nurse, M. (2004) 'Linkages between community forestry and poverty', Regional Community Forestry Training Centre for Asia and the Pacific, Bangkok

Gilmour, D., O'Brien, N. and Nurse, M. (2005) 'Overview of regulatory frameworks for community forestry in Asia', in N. O'Brien, S. Matthews and M. Nurse (eds) *First Regional Community Forestry Forum: Regulatory Frameworks for Community Forestry in Asia*. Proceedings of a Regional Forum, RECOFTC, August 24–25, Bangkok

GoI (2008) *India 2008: A Reference Manual*. Ministry of Information and Broadcasting, Government of India, New Delhi

Gómez, I. and Méndez, V. E. (2005) 'Análisis de Contexto: el Caso de la Asociación de Comunidades Forestales de Petén (ACOFOP)' (Contextual Analysis: Association of Forest Communities of the Petén), PRISMA, San Salvador, El Salvador

GoR (2007) *Economic Review 2006–07*. Directorate of Economics and Statistics, Government of Rajasthan, Jaipur, India

Griffiths, R. (2008) 'Seeing "REDD"? Forests, climate change mitigation and the rights of indigenous peoples and local communities', update for Poznan (UNFCCC COP 14), Forest Peoples Program, England and Wales

GTZ (Gesellschaft für Technische Zusammenarbeit) (2004) 'The ILO Convention 169', Indigenous Peoples in Latin America and the Caribbean, available at www2.gtz.de/indigenas/english/international-instruments/ilo169.htm (last accessed April 2009)

Guiang, E. S. and Castillo, G. (2007) 'Trends in forest ownership, forest resources, tenure and institutional arrangements in the Philippines: are they contributing to better forest management and poverty reduction?', in *Understanding Forest Tenure in South and Southeast Asia,* Forest Policy and Institutions Working Paper 14, FAO, Rome, available at www.fao.org/docrep/009/j8167e/j8167e00.htm (last accessed September 2009)

Gurung, K. (2006) 'New distillation units for sustainable management and processing of non timber forest products (NTFPs) projects, Dolakha', GEF, Small Grants Program, Dolakha, Nepal

Habermas, J. (1981) 'New social movements', *Telos*, vol 49, pp33–73

Habib, A. and Kotze, H. (2002) 'Civil society, governance and development in an era of globalisation', unpublished manuscript

Hallberg, K. (2000) 'A market-oriented strategy for small and medium scale enterprises', World Bank and International Finance Corporation, Washington, DC

Harrison, R. P. (1992) *Forests: The Shadow of Civilization*. University of Chicago Press, Chicago, Illinois, USA

Harvey, D. (2003) *The New Imperialism*. Oxford University Press, Oxford

Hasan, U., Irawan, D. and Komarudin, H. (2008) 'Rio: Modal sosial Sistem Pemerintah Desa' (Rio: The Social Capital of a Village Government), in H. Adnan, D. Tadjudin, E. L. Yuliani, H. Komarudin, D. Lopulalan, Y. L. Siagian and D. W. Munggoro (eds) *Belajar Dari Bungo: Mengelola Sumberdaya Alam di Era Desentralisasi* (Learning from Bungo: Managing Natural Resources in the Era of Decentralization), CIFOR, Bogor, Indonesia

Hayami, Y. (1998) 'Community, market and state', in C. Eicher and J. Staatz (eds) *International Agricultural Developments*. Johns Hopkins University Press, Baltimore, Maryland, USA

Herlihy, P. H., and Knapp, G. (2003) 'Maps of, by, and for the peoples of Latin America', *Human Organization*, vol 62, no 4, pp303–314

Heywood, V. H. (1995) *Global Biodiversity Assessment*. Cambridge University Press, Cambridge, UK

Hickey, S. and Bracking, S. (2005) 'Exploring the politics of chronic poverty: from representation to a politics of justice?' *World Development*, vol 33, no 6, pp851–865

HMG/MoLJ (1993) Forest Act, 1993, His Majesty's Government of Nepal and Ministry of Law and Justice, Kathmandu, Nepal

Hobley, M. (2007) 'Where in the world is there pro-poor forest policy and tenure reform?', Rights and Resources Initiative, Washington, DC

Ibarguen, R. (2008) *La última frontera y las comunidades de pequeños parcelarios en el norte paceño*. CEDLA and CIFOR, La Paz, Bolivia

INE (2001) Resultados finales del Censo Nacional de Población y Vivienda de 2001. Ministerio de Desarrollo Sostenible y Planificacion, Instituto Nacional de Estadísticas, La Paz, Bolivia

INE (2002) Censo de Población y Vivienda 2001. INE, La Paz, Bolivia

INEC (National Institute of Statistics and Census (Nicaragua)) (2005) Resumen Censal. VII Censo de Población y IV de Vivienda, www.inec.gob.ni/censos2005/ResumenCensal/Resumen2.pdf (last accessed April 2008)

Inoue, M. and Isozaki, H. (eds) (2003) *People and Forests: Policy and Local Reality in Southeast Asia, the Russian Far East and Japan*. Kluwer Academic Publishers, The Netherlands

Intelsig (2008) 'Análisis multitemporal aplicando imágenes satélite para la cuantificación de los cambios de uso de la tierra y cobertura en BOSAWAS-RAAN y en los departamentos de Rivas, Carazo y Granada', Final project report, GTZ/GFA, Managua

IPCC (Intergovernmental Panel on Climate Change) (2007) Synthesis Report, IPCC Plenary XXVII Valencia, Spain, November 12–17

IRMA (2006) 'NTGCF: "Anand Pattern" in natural resources management', Institute of Rural Management, available at www.irma.ac.in/about/ntgcf.html (last accessed November 2006)

Iversen, V., Chhetry, B., Francis, P., Gurung, M., Kafle, G., Pain, A. and Seeley, J. (2006) 'High value forests, hidden economies and elite capture: evidence from forest user groups in Nepal's Terai', *Ecological Economics*, vol 58, pp93–107

Jelin, E. (1986) 'Los movimientos sociales ante la crisis', Universidad de las Naciones Unidas, Buenos Aires

Junkin, R. (2007) 'Overcoming the barriers to financial services for small-scale forestry: the case of the community forest enterprises of Petén, Guatemala', *Unasylva*, vol 228, no 58, pp38–43

Kaimowitz, D. (2003a) 'Forest law enforcement and rural livelihoods', *International Forestry Review*, vol 5, no 3, pp199–210

Kaimowitz, D. (2003b) 'Not by bread alone...forests and rural livelihoods in sub-Saharan Africa' in T. Oksanen, B. Pajari and T. Tuomasjukka (eds) *Forests in Poverty Reduction Strategies: Capturing the Potential*. EFI Proceedings 47, Tuusula, Finland

Kanel, K. R. (2004) *Twenty Five Years of Community Forestry: Contribution to Millennium Development Goals*. Proceedings of Fourth National Workshop on Community Forestry

Kanel, K. R., Poudyal, R. P. and Baral, J. P. (2005) *Nepal: Community Forestry*. Proceedings of the First Regional Community Forestry Forum, Recoftc, Bangkok

Kante, B. (2008) *Amélioration de l'équité et des moyens de subsistance dans la foresterie communautaire au Burkina Faso*. CIFOR and RRI, Ouagadougou, Burkina Faso

Kasanga, K. and Kotey, N. A. (2001) *Land Management in Ghana: Building on Tradition and Modernity*. International Institute for Environment and Development, London

Kerkhoff, E. and Erni, C. (eds) (2005) 'Shifting cultivation and wildlife conservation: a debate', *Indigenous Affairs*, vol 2, pp22–29

Khare, A. and Bray, D. B. (2004) 'Study of the critical new forest conservation issues in the Global South', Final Summary Report, Ford Foundation

Kimerling, J. (1991) 'Disregarding environmental law: petroleum development in protected areas and indigenous homelands in the Ecuadorian Amazon', *Hastings International and Comparative Law Review*, vol 14, pp849–903

Komarudin, H., Siagian, Y. and Colfer, C. (2008) 'Collective action to secure property rights for the poor', Working Paper 90, CAPRi, Washington, DC

Koops, B.-J., Lips, M., Prins, C. and Schellekens, M. (2006) *Starting Points for ICT Regulations: Deconstructing Prevalent Policy One-liners*. Cambridge University Press, Cambridge

Kozak, R. (2007) *Small and Medium Forest Enterprises: Instruments of Change in the Developing World*. Rights and Resources Initiative and University of British Columbia, Washington, DC

Kuechli, C. and Blaser, J. (2005) 'Forests and decentralization in Switzerland: a sampling', in C. J. P. Colfer and D. Capistrano (eds) *The Politics of Decentralization*. Earthscan, London

Kull, C.A. (2004) *Isle of Fire: The Political Ecology of Landscape Burning in Madagascar*. University of Chicago Press, Chicago, Illinois, USA

Larson, A. M. (2008) 'Land tenure rights and limits to forest management in Nicaragua's North Atlantic Autonomous Region: making the rules of the game', paper presented at conference of International Association for the Study of the Commons (IASC), July 14–18, Cheltenham, England

Larson, A. M. and Ribot, J. C. (2007) 'The poverty of forestry policy: double standards on an uneven playing field', *Sustainability Science*, vol 2, no 2, pp189–204

Larson, A. M. and Soto, F. (2008) 'Decentralization of natural resource governance regimes', *Annual Review of Environment and Resources*, vol 33, pp213–239

Larson, A. M. and Mendoza-Lewis, J. (2009) *Desafíos en la Tenencia Comunitaria de Bosques en la RAAN de Nicaragua*. CIFOR/URACCAN/RRI, Managua

Larson, A. M., Cronkleton, P., Barry, D. and Pacheco, P. (2008) 'Tenure rights and beyond: community access to forest resources in Latin America', Occasional Paper 50, CIFOR, Bogor, Indonesia

Leyva, S., Burgeuete, A. and Speed, S. (eds) (2008) *Gobernar en la Diversidad: Experiencias Indígenas Desde América Latina*. Centro de Investigaciones y Estudios Superiores en Antropología Social, Facultad Latinoamericana de Ciencias Sociales

Li, T. M. (2002) 'Engaging simplifications: community-based resource management, market processes and state agendas in upland Southeast Asia', *World Development*, vol 30, no 2, pp265–283

López, G. R. (2004) 'Negociaron tierras fiscales en la TCO de Guarayos', *El Deber*, 7 November, Santa Cruz de la Sierra, Bolivia

Luintel, H. (2002) 'Issues and options of sustainable management of Himalayan medicinal herbs', *Journal of Forest and Livelihood*, vol 2, no 1, pp53–55

Lynch, O. J. and Harwell, E. (eds) (2002) *Whose Natural Resources? Whose Common Good? Towards a New Paradigm of Environmental Justice and the National Interest in Indonesia*. ELSAM, Lembaga Studi dan Advokasi Masyarakat (Institute for Policy Research and Advocacy), Jakarta

Macchi, M., Oviedo, G., Gotheil, S., Cross, K., Boedhihartono, A., Wolfangel, C. and Howell, M. (2008) *Indigenous and Traditional Peoples and Climate Change*. IUCN, Gland, Switzerland

MacPherson, C. B. (ed.) (1978) *Property: Mainstream and Critical Positions*. University of Toronto Press, Toronto, Ontario, Canada

Macqueen, D. (2008) 'Small and medium forestry enterprise: supporting small forest enterprises', IIED, London

Magno, F. (2001) 'Forest devolution and social capital: state-civil society relations in the Philippines', *Environmental History*, vol 6, no 2, pp264–286

Maisels, F., Sunderland, T., Curran, B., von Liebenstein, K., Oates, J., Usongo, L., Dunn, A., Asaha, S., Balinga, M., Defo, L. and Telfer, P. (2007) 'Central Africa's protected areas and the purported displacement of people: a first critical review of existing data', in K. Redford and E. Fearn (eds) 'Protected Areas and Human Displacement: A Conservation Perspective', Working Paper 27, Wildlife Conservation Society

Malla, Y. (2000) 'Impact of community forestry policy on rural livelihoods and food security in Nepal', *Unasylva*, vol 51, no 3

Mamdani, M. (1996) *Citizen and Subject: Contemporary Africa and the Legacy of Late Colonialism*. Princeton University Press, Princeton, New Jersey, USA, and David Phillip, Cape Town

Mantel, K. (1964) 'History of the international science of forestry with special consideration of Central Europe: literature, training, and research from the earliest beginnings to the nineteenth century', in J. A. Romberger and P. Mikola (eds) *International Review of Forestry Research*, vol 1, pp1–37, Academic Press, New York

Marfo, E. (2001) 'Community interest representation in negotiation: a case of the social responsibility agreement in Ghana', MSc thesis, Wageningen University, The Netherlands

Marfo, E. (2004) 'Unpacking and repacking community representation in forest policy and management negotiations: lessons from the social responsibility agreement in Ghana', *Ghana Journal of Forestry*, vol 15–16, pp20–29

Marfo, E. (2006) 'Powerful relations: the role of actor-empowerment in the management of natural resource conflicts. A case of forest conflicts in Ghana', PhD thesis (published), Wageningen University, The Netherlands

Marfo, E. (2009) *Security of Tenure Reforms and Community Benefits Under Collaborative Forest Management Arrangements in Ghana: A country report*. CIFOR and RRI, Accra, Ghana

McCarthy, J. D. and Zald, M. N. (eds) (1977) *The Dynamics of Social Movements.* Winthrop Publishers, Massachusetts, USA

McCarthy, J., Barr, C., Resosudarmo, I. A. P. and Dermawan, A. (2006) 'Origins and scope of Indonesia's decentralization laws', in C. Barr, I. A. P. Resosudarmo, A. Dermawan and J. McCarthy, with M. Moeliono and B. Setiono (eds) *Decentralization of Forest Administration in Indonesia: Implications for Forest Sustainability, Economic Development and Community Livelihoods.* CIFOR, Bogor, Indonesia

McDermott, M. H. (2001) 'Invoking community: indigenous people and ancestral domain in Palawan, the Philippines', in A. Agrawal and C. Gibson (eds) *Communities and the Environment: Ethnicity, Gender and the State in Community-Based Conservation.* Rutgers University Press, New Brunswick, New Jersey, USA

Meinzen-Dick, D. (2006) 'Shifting boundaries of tenure systems and security of access to common property', paper presented at 11th Biennial Conference International Association for the Study of Common Property, Ubud, Bali

Meinzen-Dick, R. and Pradhan, R. (2001) 'Implications of legal pluralism for natural resource management', in L. Mehta, M. Leach and I. Scoones (eds) Environmental Governance in an Uncertain World, *IDS Bulletin*, vol 32, no 4, pp10–17

Meinzen-Dick, R. and Mwangi, E. (2008) 'Cutting the web of interests: pitfalls of formalizing property rights', *Land Use Policy*, vol 26, pp36–43

Misra, V. K. (2002) 'Greening of wastelands: experiences from the Tree Growers' Cooperative Project', in D. K. Marothia (ed.) *Institutionalizing Common Pool Resources.* Concept Publishing, New Delhi

Mollinedo, A. C., Campos, J. J., Kanninen, M. and Gómez, M. (2002) *Beneficios sociales y rentabilidad financiera del manejo forestal comunitario en la Reserva de la Biósfera Maya, Guatemala.* CATIE, Serie Técnica, Informe Tecnico No. 327

Molnar, A. (2003) *Forest Certification and Communities: Looking Forward to the Next Decade.* Forest Trends, Washington, DC

Molnar, A., Scherr, S. and Khare, A. (2004) *Who Conserves the World's Forests? Community-Driven Strategies to Protect Forests and Respect Rights.* Forest Trends, Washington, DC

Molnar, A., Liddle, M., Bracer, C., Khare, A., White, A. and Bull, J. (2007) 'Community-based forest enterprises in tropical forest countries: status and potential', International Tropical Timber Organization, Rights and Resources Initiative and Forest Trends, Washington, DC

Mongbo, R. (2008) 'State building and local democracy in Benin: two cases of decentralized forest management', *Conservation and Society*, vol 6, no 1, pp49–61

Monterroso, I. (2007) 'Extracción de xate en la Reserva de Biósfera Maya: Elementos para una evaluación de su sostenibilidad', Editorial FLACSO-Guatemala

Monterroso, I. and Barry, D. (2007) 'Community-based forestry and the changes in tenure and access rights in the Mayan Biosphere Reserve, Guatemala', paper presented at International Conference on Poverty Reduction and Forests: Tenure, Market & Policy Reforms, September 3–7, RECOFTC, Bangkok

Monterroso, I. and Barry, D. (2008) *Sistema de Concesiones Forestales Comunitarias: Tenencia de la Tierra, Bosques y Medios de Vida en la Reserva de la Biósfera Maya en Guatemala.* CIFOR and FLACSO, Guatemala City, Guatemala

Monterroso, I. and Barry, D. (2009) *Sistema de Concesiones Forestales Comunitarias: reflexiones sobre la reforma forestal y el futuro del modelo.* CIFOR and FLACSO, Guatemala

MoRD and NRSA (2005) *Wastelands Atlas of India.* Ministry of Rural Development, Government of India, New Delhi, and National Remote Sensing Agency, Hyderabad

Moreira, E. and Hébette, H. (2003) 'Estudo socio-economico com vista a criacao da Resex Verde para Sempre', UFPA, Belem, Brazil

Moreno, R. D. (2006) 'COPNAG denuncia la venta en $us 1,2 millones de TCO en Guarayos', in *El Deber*, 27 November, Santa Cruz de la Sierra, Bolivia

Mwangi, E. and Dohrn, S. (2008) 'Securing access to dryland resources for multiple users in Africa: a review of recent research', *Land Use Policy*, vol 25, pp240–248

Navarro, G., Del Gatto, F., Faurby, O. and Arguello, A. (2007) 'Verificación de la Legalidad en el Sector Forestal Nicaragüense', VI Congreso Forestal Centroamericano 'Competitividad, Sostenibilidad Forestal en Centroamérica' ('Verification of Legality in the Nicaraguan Forest Sector', VI Central American Forestry Congress, 'Competitiveness, Forest Sustainability in Central America'), August 29–31, San Salvador

Navarro, G., Sánchez, M., Larson, A., Bermúdez, G. and Méndez E. (2008) 'Simplificación de trámites en el sistema de verificación de la legalidad del sector forestal en Nicaragua' (Simplification of paperwork in the system for the verification of legality in Nicaragua's forest sector), unpublished consultancy report, Informe 2: Diagnóstico de los Permisos Forestales de Aprovechamiento. Instituto Nacional Forestal (INAFOR)/Deutsche Gesellschaft für Technische Zusammenarbeit (GTZ)/ Centro Agronómico Tropical de Investigación y Enseñanza (CATIE), Managua

Neidhardt, F. and Rucht, D. (1991) *The Analysis of Social Movements: The State of the Art and Some Perspectives for Further Research*. Westview Press, Boulder, Colorado, USA

Nightingale, A. (2002) 'Participating or just sitting in? The dynamics of gender and caste in community forestry', *Journal of Forestry and Livelihoods*, vol 2, no 1, pp17–24

Nittler, J. and Tschinkel, H. (2005) 'Manejo comunitario del bosque en la RBM de Guatemala: Protección mediante ganancias', Collaborative Research Support Program 32, Sustainable Agriculture and Natural Resources Management, University of Georgia, Watkinsville, Georgia, USA

NTGCF (1997) *Annual Report 1996-97*. National Tree Growers' Cooperative Federation Limited, Anand, Gujarat, India

Ntsebeza, L. (2005) 'Democratic decentralization and traditional authority: dilemmas of land administration in rural Africa', in J. C. Ribot and A. M. Larson (eds) *Democratic Decentralization through a Natural Resource Lens*. Routledge, London

Nunes, W., Mourão, P., Lobo, R. and Cayres, G. (2008) *Entre sonhos e pesadelos: acesso a terra e manejo florestal nas comunidades rurais em Porto de Moz*. CIFOR, Belem, Brazil

O'Brien, N., Matthews, S. and Nurse, M. (eds) (2005) 'Regulatory frameworks for community forestry in Asia', First Regional Community Forestry Forum, Proceedings of a Regional Forum, RECOFTC, Bangkok, Thailand, pp3–33

Offe, C. (1985) 'New social movements: challenging the boundaries of institutional politics', *Social Research*, vol 52, no 4, pp817–868

Ojha, H., Khanal, D. R., Paudel, N. S., Sharma, H. and Pathak, B. (2007) 'Federation of community forest user groups in Nepal: an innovation in democratic forest governance', Proceedings of International Conference on Poverty Reduction and Forests', RECOFTC and RRI, September, Bangkok

Ojha, H. R., Timsina, N. P., Chhetri, R. and Paudel, K. (2008) *Communities, Forests and Governance: Policy and Institutional Innovations from Nepal*. Adroit Publishers, New Delhi, India

Onibon, A., Dabiré, B. and Ferroukhi, L. (1999) 'Local practice and decentralization and devolution of natural resources management in West Africa', *Unasylva*, vol 50, pp23–27

Opoku, K. (2006) 'Forest governance in Ghana: an NGO perspective: a report produced for FERN', Forest Watch, Ghana

Ostrom, E. (1990) *Governing the Commons: The Evolution of Institutions for Collective Action*. Cambridge University Press, New York

Ostrom, E. (1999) 'Self-governance and forest resources', Occasional Paper 20, CIFOR, Bogor, Indonesia

Ostrom, E. (2000) 'El gobierno de los comunes: la evolución de las instituciones de acción colectiva', Editorial Fondo de Cultura Económica, México

Otsuka, K. and Place, F. (eds) (2001) *Land Tenure and Natural Resource Management: A Comparative Study of Agrarian Communities in Asia and Africa*. Johns Hopkins University Press, Washington, DC

Oviedo, G. (2002) 'Lessons learned in the establishment and management of protected areas by indigenous and local communities', mimeo, World Conservation Union (IUCN), Washington, DC

Owusu, M. (1996) 'Tradition and transformation: democracy and the politics of popular power in Ghana', *Journal of Modern Africa Studies*, vol 34, no 2, pp307–343

Oyono P. R. (2002) 'Forest management, systemic crisis and policy change: socio-organizational roots of ecological uncertainties in the Cameroon's decentralization model', paper presented at World Resources Institute Conference on 'Decentralization and the Environment', Bellagio, Italy

Oyono, P. R. (2004a) 'One step forward, two steps back? Paradoxes of natural resources management decentralisation in Cameroon', *Journal of Modern African Studies*, vol 42, no 1, pp91–111

Oyono, P. R. (2004b) 'Institutional deficit, representation, and decentralized forest management in Cameroon', Environmental Governance in Africa Working Paper 15, World Resources Institute, Washington, DC

Oyono, P. R. (2005a) 'Profiling local level outcomes of environmental decentralizations: the case of Cameroon's forests in the Congo Basin', *Journal of Environment and Development*, vol 14, no 2, pp1–21

Oyono, P. R. (2005b) 'Social and organizational roots of ecological uncertainties in Cameroon's forest management decentralization model', in J. C. Ribot and A. M. Larson (eds) *Democratic Decentralization through a Natural Resource Lens*. Routledge, London

Oyono, P. R., Ribot, J. C. and Larson, A. M. (2006) 'Green and black gold in rural Cameroon: natural resources for local justice, governance and sustainability', Working Paper 22, World Resources Institute, Washington, DC

Oyono, P. R., Kombo, S. S. and Biyong, M. B. (2008) 'New niches of community rights to forest in Cameroon: cumulative effects on livelihoods and local forms of vulnerability', Country Synthesis Report, CIFOR, Yaounde, Cameroon

Pacheco, P. (2006) 'Acceso y uso de la tierra y bosques en Bolivia: sus implicaciones para el desarrollo y la conservación', UDAPE, La Paz, Bolivia

Pacheco, P., Barry, D., Cronkleton, P. and Larson, A. (2008a) 'From agrarian to forest tenure reforms in Latin America: assessing their impacts for local people and forests', paper presented at Conference of the International Association for the Study of the Commons (IASC), July 14–18, Cheltenham, England

Pacheco, P., Barry, D., Cronkleton, P. and Larson, A. (2008b) 'The role of informal institutions in the use of forest resources in Latin America', Forests and Governance Program Paper 15, CIFOR, Bogor, Indonesia

Pacheco, P., Ibarra, E., Cronkleton, P. and Amaral, P. (2008c) 'Políticas Públicas que Afectan el Manejo Forestal Comunitario', in C. Sabogal, B. Pokorny, W. de Jong, B. Louman, P. Pacheco, D. Stoian and N. Porro (eds) *Manejo forestal comunitario en*

América Tropical: Experiencias, lecciones aprendidas y retos para el futuro. CIFOR, Bogor, Indonesia

Pagdee, A., Kim, Y. and Daugherty, P. J. (2006) 'What makes community forest management successful: a meta study from community forests throughout the world', *Society and Natural Resources*, vol 19, no 1, pp33–52

Paudel, N. and Banjade, M. (2008a) 'Sundari Community Forest Users Group, Dolakha', unpublished draft, Forest Action, Kathmandu, Nepal

Paudel, N. S. and Banjade, M. R. (2008b) 'Improving equity and livelihoods in community forestry: Sundari Community Forest Users Group', Nawalparasi site report, CIFOR, Forest Action, Kathmandu, Nepal

Paudel, D., Keeling, S. J. and Khanal, D. R. (2006) 'Forest products verification in Nepal and the work of the Commission to investigate the abuse of authority', VERIFOR Case Study 10, available at www.verifor.org (last accessed September 2009)

Paudel, N. S., Banjade, M. and Dahal, G. (2008a) 'Improving equity and livelihoods in community forestry, Country Report Nepal', Forest Action and CIFOR, Kathmandu

Paudel, N.S., Banjade, M. R. and Dahal, G. (2008b) 'Community forestry in changing context: changing livelihoods and emerging market opportunities', ForestAction and CIFOR, Kathmandu

Peluso, N. L. (1990) 'A history of state forest management in Java', in M. Poffenberger (ed.) *Keepers of the Forest: Land Management Alternatives in Southeast Asia*. Kumarian Press, Hartford, Connecticut, USA

Peluso, N. L. (1992) *Rich Forests, Poor People: Resource Control and Resistance in Java*. University of California Press, Berkeley, California, USA

Peluso, N. L. (1994) *The Impact of Social and Environmental Change on Forest Management: A Case Study from West Kalimantan, Indonesia*, vol 8, FAO Community Forestry Case Study Series, Food and Agriculture Organization, Rome

Plant, R. and Hvalkof, S. (2001) 'Land titling and indigenous peoples', Sustainable Development Department Technical Papers Series, Inter-American Development Bank, Washington, DC

Poffenberger, M. (ed.) (1990) *Keepers of the Forest: Land Management Alternatives in Southeast Asia*. Kumarian Press, West Hartford, Connecticut, USA

Poffenberger, M. (1996) 'Grassroots forest protection: Eastern India experiences', Research Network Report 7, Center for Southeast Asian Studies, University of California–Berkeley, USA

Poffenberger, M., Walpole, P., D'Silva, E., Lawrence, K. and Khare, A. (1997) 'Linking government with community resource management: what's working and what's not', Research Network Report 9, Center for Southeast Asian Studies, University of California–Berkeley, USA

Pokorny, B. and Johnson, J. (2008) 'Community forestry in the Amazon: the unsolved challenge of forests and the poor', *Natural Resources Perspective*, vol 112, Overseas Development Institute, London

Pulhin, J. M. (2006) 'People, power and timber: politics of resource use in community-based forest management', Vicente Lu Professional Chair Lecture in Forestry, College of Forestry and Natural Resources, University of the Philippines at Los Banos, Laguna

Pulhin, J. M. and Dizon, J. T. (2003) 'Politics of tenure reform in the Philippine forest land', Politics of the Commons: Articulating Development and Strengthening Local Practices, July 11–14, Chiang Mai, Thailand

Pulhin, J. M. and Ramirez, M. A. (2006) 'Behind the fragile enterprise: community-based timber utilization in southern Philippines', final report of case study submitted to

Regional Community Forestry Training Center for Asia and the Pacific (RECOFTC) with the financial assistance of International Tropical Timber Organization (ITTO) and Forest Trends

Pulhin, J. M. and Ramirez, M. A. (2008) 'Improving equity and livelihoods in community forestry: the case of Ngan, Panansalan, Pagsabangan Forest Resources Development Cooperative, Inc (NPPFRDC), Compostela Valley, Southern Philippines', CIFOR, Manila

Pulhin, J. M., Inoue, M. and Enters, T. (2007) 'Three decades of community-based forest management in the Philippines: emerging lessons for sustainable and equitable forest management', *International Forestry Review*, vol 19, no 4, pp865–883

Pulhin, J. M., Dizon, J. T., Cruz, R. V. O., Gevaña, D. T. and Dahal, G. R. (2008) 'Tenure reform on Philippine forest lands: assessment of socio-economic and environmental impacts', College of Forestry and Natural Resources, University of Philippines Los Banos

Quijandría, B., Monares, A. and Ugarte, R. (2001) *Assessment of Rural Poverty: Latin America and the Caribbean*. IFAD, Santiago, Chile

Raharjo, D. Y., Oktavia, V. and Azmaiyanti, Y. (2004) *Obrolan Lapau, Obrolan Rakyat: Sebuah Potret Pergulatan Kembali ke Nagari (Cafe Conversation, the People's Conversation: A Portrait of the Struggle to Return to Nagari [the customary/traditional governance system of the Minangkabau of West Sumatra])*. Studio Kendil, Bogor, Indonesia

Ranjatson, J. P. (2009) 'La Gouvernance Nationale du Processus Koloala et Quelques Implications pour le Projet KAM', ESSA-Forêts, Univesité d'Antananarivo, Antananarivo, Madagascar

Redford, K. H., Coppolillo, P., Sanderson, E. W, Fonseca, G. A. B., Groves, C., Mace, G., Maginnis, S., Mittermier, R., Noss, R., Olson, D., Robinson, J.G., Vedder, A. and Wright, M. (2003) 'Mapping the conservation landscape', *Conservation Biology*, 17 (1), pp116–132

Resosudarmo, I. A. P. (2005) 'Closer to people and trees: will decentralization work for the people and forests of Indonesia?', in J. C. Ribot and A. M. Larson (eds) *Democratic Decentralization through a Natural Resource Lens*. Routledge, London

Ribot, J. (1999) 'Decentralization, participation and accountability in Sahelian forestry: legal instruments of policial-administrative control', *Africa*, vol 69, no 1, pp23–65

Ribot, J. (2002) 'African decentralization: local actors, powers and accountability', Democracy, Governance and Human Rights Paper 8, UNRISD and IDRC, Geneva

Ribot, J. (2004) 'Waiting for democracy: the politics of choice in natural resource decentralization', World Resources Institute, Washington, DC

Ribot, J. C. and Peluso, N. L. (2003) 'A theory of access', *Rural Sociology*, vol 68, no 2, pp153–181

Ribot, J. C. and Larson, A. M. (eds) (2005) *Democratic Decentralisation through a Natural Resource Lens*. Routledge, London

Ribot, J. C., Chhatre, A. and Lankina, T. (2008) 'Introduction: institutional choice and recognition in the formation and consolidation of local democracy', *Conservation and Society*, vol 6, no 1, pp1–11

Rice, D. (1994) 'Clearing our own Ikalahan path', in J. B. Raintree and H. A. Francisco (eds) *Marketing of Multipurpose Tree Products in Asia*. Conference Proceedings of Multipurpose Tree Species Research Network in Asia, available at www.fao.org/docrep/x0271e/x0271e03.htm (last accessed January 2009)

Rice, D. (2001) *Forest Management by a Forest Community: The Experience of the Ikalahan*. Kalahan Educational Foundation, Inc.

Ritchie, B., McDougall, C., Haggith, M. and Burford de Oliveira, N. (2000) *An Introductory Guide to Criteria and Indicators for Sustainability in Community Managed Forest Landscapes*. CIFOR, Bogor, Indonesia

Roe, D. (2008) 'The origins and evolution of the conservation-poverty debate: a review of key literature, events and policy processes', *Oryx*, vol 42, no 4, pp491–503

Roldan, R. (2004) 'Models for recognizing indigenous land rights in Latin America', Biodiversity Series Paper 99, Environment Department, World Bank, Washington, DC

Romano, F. (2007) 'Forest tenure change in Africa: making locally based forest management work', *Unasylva*, vol 228, no 57, pp11–17

Rosset, P., Patel, R. and Courville, M. (eds) (2006) 'Promised land: competing visions of agrarian reform', Land Research Action Network

RRI (2009) *Who Owns the Forests of Asia? An Introduction to the Forest Tenure Transition in Asia, 2002–2008*. Rights and Resources Initiative, Washington, DC

Ruiz, S. (2005) *Rentismo, conflicto y bosques en el norte amazónico boliviano*. CIFOR, Santa Cruz, Bolivia

Saigal, S., Dahal, G. R. and Vira, B. (2008) 'Cooperation in forestry: analysis of forestry cooperatives in Rajasthan, India', unpublished project report, CIFOR and RRI

Salas, H. C. (1995) Libro de lecturas del taller sobre reforma de las políticas de gobierno relacionadas con la conservación y el desarrollo forestal en América Latina, 1–3 de junio 1994, Washington, DC, IICA Biblioteca Venezuela

Salgado, I. and Kaimowitz, D. (2003) 'Porto de Moz: O prefeito "dono do município"', in F. Toni (ed.) *Municípios e Gestão Florestal na Amazônia*. A.S. Editores, Brasília

Sarin, M., Singh, N., Sundar, N. and Bhogal, R. (2003) 'Devolution as a threat to democratic decision-making in forestry? Findings from three states in India', in D. Edmunds and E. Wollenburg (eds) *Local Forest Management: The Impacts of Devolution Policies*. Earthscan, London

Sasu, O. (2005) 'Decentralization of federal forestry systems in Ghana', in C. J. P. Colfer and D. Capistrano (eds) *The Politics of Decentralization*. Earthscan, London

Saxena, R. (1996) 'The Vatra Tree Growers' Cooperative Society', in K. Singh and V. Ballabh (eds) *Cooperative Management of Natural Resources*. Sage Publications, New Delhi

Sayer, J., McNeely, J., Maginnis, S., Boedhihartono, I., Shepherd, G. and Fisher, B. (2008) *Local Rights and Tenure for Forests Opportunity or Threat for Conservation?* Rights and Resources Initiative and IUCN, Washington, DC

Scherr, S., White, A. and Kaimowitz, D. (2002) 'Making markets work for forest communities', Policy Brief, CIFOR, Bogor, Indonesia, and Forest Trends, Washington, DC

Scherr, S. J., White, A. and Kaimowitz, D. (2003) 'Making markets work for forest communities', *International Forestry Review*, vol 5, no 1, pp67–73

Scherr, S. J., White, A. and Kaimowitz, D. (2004) 'A new agenda for forest conservation and poverty reduction: making markets work for low-income producers', Forest Trends, CIFOR and IUCN, Washington, DC

Schlager, E. and Ostrom, E. (1992) 'Property rights regimes and natural resources: a conceptual analysis', *Land Economics*, vol 68 , no 3, pp249–62

Schmink, M. and Wood, C. H. (1984) *Frontier Expansion in Amazonia*. University of Florida Press, Gainesville, Florida, USA

Schmink, M. and Wood, C. H. (1992) *Contested Frontiers in Amazonia*. Columbia University Press, New York

Schroeder, R. A. (1999) 'Community, forestry and conditionality in the Gambia', *Africa*, vol 69, pp1–22

Science (2008) 'Special Issue: Forests in Flux', 13 June, vol 320, no 5882, p1435–1462

Scott, J. C. (1995) 'State simplifications: nature, space and people', *Journal of Political Philosophy*, vol 3, no 3, pp191–233

Seymour, F. (2008) 'Forests, climate change, and human rights: managing risk and trade-offs', in S. Humphreys (ed.) *Human Rights and Climate Change*. International Council on Human Rights Policy, Cambridge University Press, Cambridge

Sikor, T. and Thanh, T. N. (2007) 'Exclusive versus inclusive devolution in forest management: insights from forest land allocation in Vietnam's Central Highlands', *Land Use Policy*, vol 24, pp644–653

Sikor, T. and Lund, C. (2009) 'Access and property: a question of power and authority', *Development and Change*, vol 40, no 1, pp1–22

Singh, B. P. (2007) Personal communication with senior project officer of Foundation for Ecological Security (FES), April 17

Singleton, S. (1998) *Constructing Cooperation: The Evolution of Institutions of Comanagement*. University of Michigan Press, Ann Arbor, Michigan, USA

Smith, W. (2006) 'Regulating timber commodity chains: timber commodity chains linking Cameroon and Europe', paper presented at conference of the International Association for the Study of Common Property (IASCP), Bali, Indonesia

Solares, A. M. (2008) 'Las MIPYMES en las exportaciones bolivianas', USAID and IBCE, La Paz, Bolivia

Spierenburg, M., Steenkamp, C. and Wels, H. (2008) 'Enclosing the local for the global commons: community land rights in the Great Limpopo Transfrontier Conservation Area', *Conservation and Society*, vol 6, no 1, pp87–97

Stocks, A. (2005) 'Too much for too few: problems of indigenous land rights in Latin America', *Annual Review of Anthropology*, vol 34, pp85–104

Stocks, A., McMahan, B. and Taber, P. (2007) 'Indigenous, colonist, and government impacts on Nicaragua's Bosawas reserve', *Conservation Biology*, vol 21, no 6, pp1495–1505

Stoian, D. (2000) 'Variations and dynamics of extractive economies: the rural urban nexus of non-timber forest use in the Bolivian Amazon', PhD thesis, University of Freiburg at Freiburg, Germany

Stoian, D. (2004) 'Cosechando lo que cae: La economía de la castaña (*Bertholletia excelsa* H.B.K.) en la Amazonía Boliviana', in M. Alexiades and P. Shanley (eds) *Productos Forestales, Medios de Subsistencia y Conservación*, vol 3, América Latina, CIFOR, Bogor, Indonesia

Stoian, D. (2005) 'Making the best of two worlds: rural and peri-urban livelihood options sustained by non-timber forest products from the Bolivian Amazon', *World Development*, vol 33, no 9, pp1473–1490

Stoian, D. and Henkemans, A. (2000) 'Between extractivism and peasant agriculture: differentiation of rural settlement in the Bolivian Amazon', *International Tree Crops Journal*, vol 10, pp299–319

Subedi, B. P. (2006) *Linking Plant-Based Enterprises and Local Communities to Biodiversity Conservation in Nepal Himalaya*. Adroit Publishers, New Delhi

Sundberg, J. (1998) 'NGO landscapes in the Mayan Biosphere Reserve, Guatemala', *Geographical Review*, vol 88, no 3, pp388–412

Sunderlin, W. D. and Pokam, J. (2002) 'Economic crisis and forest cover change in Cameroon: the roles of migration, crop diversification, and gender division of labor', *Economic Development and Cultural Change*, vol 50, no 3, pp581–606

Sunderlin, W. D., Angelsen, A., Belcher, B., Burgers, P., Nasi, R., Santoso, L. and Wunder, S. (2005) 'Livelihoods, forests, and conservation in developing countries: an overview', *World Development*, vol 33, pp1383–1402

Sunderlin, W., Hatcher, J. and Liddle, M. (2008) *From Exclusion to Ownership? Challenges and Opportunities in Advancing Forest Tenure Reform*. Rights and Resource Initiative, Washington, DC

Sushil, S., Dahal, G. R. and Vira, B. (2008) 'Cooperation in forestry: analysis of forestry cooperatives in Rajasthan, India', Country Synthesis Report for India, CIFOR-RRI Project, Bogor, Indonesia

Swiderska, K., with Roe, D., Siegele, L. and Grieg-Gran, M. (2009) *The Governance of Nature and the Nature of Governance: Policy That Works for Biodiversity and Livelihoods*. IIED, London

Tacconi, L. (2007a) 'Decentralization, forest and livelihoods: theory and narrative', *Global Environmental Change*, vol 12, pp338–348

Tacconi, L. (2007b) *Illegal Logging: Law Enforcement, Livelihoods and the Timber Trade*. Earthscan, London

Tahamana, B. Z. (2007) 'Understanding legal pluralism: past to present, local to global', Legal Studies Research Paper Series 07-0080, St. John's University School of Law, Queens, New York

Taylor, C. (1994) *Multiculturalism: Examining the Politics of Recognition*. Princeton University Press, Princeton, New Jersey, USA

Taylor, P., Larson, A. and Stone, S. (2007) *Forest Tenure and Poverty in Latin America: A Preliminary Scoping Exercise*. CIFOR and RRI, Bogor, Indonesia

Thiesenhusen, W. C. (1995) *Broken Promises: Agrarian Reform and the Latin American Campesino*. Westview Press, Boulder, Colorado, USA

Tilly, C. (1978) *From Mobilization to Revolution*. Addison-Wesley, Massachusetts, USA

Timsina, N. (2003) 'Viewing FECOFUN from the perspective of popular participation and representation', *Journal of Forests and Livelihoods*, vol 2, no 2, pp67–71

Touraine, A. (1985) 'An introduction to the study of social movements', *Social Research*, vol 52, no 4, pp748–787

Trópico Verde (2005) 'El proyecto turístico Cuenca del Mirador y las concesiones forestales en la zona de uso múltiple de la Reserva de la Biósfera Maya' (The Mirador Basin Tourism Project and the forest concessions in the multi-use zone of the Maya Biosphere Reserve), Trópico Verde, Flores, Guatemala

Utting, P. (2000) 'An overview of the potential and pitfalls of participatory conservation', in P. Utting (ed.) *Forest Policy and Politics in the Philippines: The Dynamics of Participatory Conservation*. Quezon City and Manila, Ateneo de Manila University Press and United Nations Research Institute for Social Development

VAIPO (Viceministerio de Asuntos Indígenas y Pueblos Originarios) (1999) 'Identificación de Necesidades Espaciales TCO Guaraya', VAIPO, La Paz, Bolivia

Vallejos, C. (1998) 'Ascensión de Guarayos: indígenas y madereros', in P. Pacheco and D. Kaimowitz (eds) *Municipios y Gestión Forestal en el Trópico Boliviano*, CIFOR, CEDLA and TIERRA, La Paz, Bolivia

van Noordwijk, M., Mulyoutami, E., Sakuntaladewi, N. and Agus, F. (2008) *Swiddens in Transition: Shifted Perceptions on Shifting Cultivators in Indonesia*. World Agroforestry Centre, Bogor, Indonesia

Vandergeest, P. and Peluso, N. (1995) 'Territorialization and state power in Thailand', *Theory and Society*, vol 24, pp385–426

Vanderlinden, J. (1989) 'Return to legal pluralism: twenty years later', *Journal of Legal Pluralism*, vol 28, pp149–157

von Benda-Beckman, F. (1997) 'Citizens, strangers and indigenous peoples: conceptual politics and legal pluralism', *Law and Anthropology*, vol 9, pp1–10

von Benda-Beckmann, C. E. and von Benda-Beckmann, F. (2002) 'Anthropology of law and the study of folk law in The Netherlands after 1950', in H. Vermeulen and J. Kommers (eds) *Tales from Academia: History of Anthropology in The Netherlands*. Verlag fur Entwicklungspolitik, Saarbrucken

von Benda-Beckman, F., von Benda-Beckman, K. and Wiber, M. (2006) 'The properties of property', in F. von Benda-Beckman, K. von Benda-Beckman and M. Wiber (eds) *Changing Properties of Property*. Berghahn, New York

von Benda-Beckmann, K. (1981) 'Forum shopping and shopping forums', *Journal of Legal Pluralism*, vol 19, pp117–159

Watts, M. and Goodman, D. (1997) 'Agrarian questions: global appetite, local metabolism: nature, culture, and industry in *fin-de-siècle* agro-food systems', in D. Goodman and M. Watts (eds) *Globalising Food: Agrarian Questions and Global Restructuring*. Routledge, New York

Weber, M. (1968) *Economy and Society: Outline of an Interpretive Sociology*. University of California Press, Berkeley, California, USA

White, A. and Martin, A. (2002) *Who Owns the World's Forests?* Forest Trends, Washington, DC

Wiggins, A. (2002) 'El caso de Awas Tingni: O el futuro de los derechos territoriales de los pueblos indígenas del Caribe Nicaragüense', reprinted in A. Rivas and R. Broegaard (eds) (2006) *Demarcación territorial de la propiedad communal en la Costa Caribe de Nicaragua*. MultiGrafic, Managua

Wollenberg, E. and Kartodihardjo, H. (2002) 'Devolution and Indonesia's new forestry law', in C. J. P. Colfer and I. A. P. Resosudarmo (eds) *Which Way Forward? People, Forests and Policymaking in Indonesia*. CIFOR and Resources for the Future, Washington, DC

Woodman, R. G. (1996) *Customary Land Law in the Ghanaian Courts*. Ghana University Press, Accra, Ghana

World Bank (2003) 'Land policies for growth and poverty reduction', Policy Research Report, Oxford University Press and World Bank, New York

Contributors

Deborah Barry is Senior Associate at the Center for International Forestry Research and Director of Country Programs for the Rights and Resources Initiative, based in Washington, DC. An economic and cultural geographer, her recent areas of work have been on community forestry in Mexico and Central America, forest tenure and governance and payment for environmental services with a concern for equity.

Carol J. Pierce Colfer, an anthropologist, is Senior Associate at the Center for International Forestry Research. In recent years her work has focused on adaptive collaborative management of forests, devolution and decentralization in forests and landscape-level forest governance. She holds a PhD in cultural anthropology from the University of Washington in Seattle, 1974, and an MPH (master of public health) in international health from the University of Hawaii, Honolulu, 1979.

Peter Cronkleton, an anthropologist with CIFOR, is a specialist in community forestry development, forest social movements and participatory research approaches. A graduate of the University of Florida (master's 1993, PhD 1998), he is currently based in Bolivia and has worked as a researcher and development practitioner in Latin America for more than 15 years.

Ganga Ram Dahal, Nepalese citizen, is Research Consultant at the Center for International Forestry Research. He obtained his PhD in forest policy and governance from the University of Reading, UK. His main areas of work include decentralization, forest tenure, community forestry and institutions.

Silvel Elías, Guatemalan citizen, has a master's and PhD in social geography from the University of Toulouse in Mirail, France. He is currently Professor at the Universidad de San Carlos in Guatemala, where he coordinates the master's in rural development and the Rural and Territorial Studies Program (PERT). His research and publications have focused on collective natural resources management and indigenous territorial rights.

Bocar Kante, citizen of Burkina Faso, is Associate Professional Officer at the Center for International Forestry Research in Burkina Faso. He is a PhD law student in Paris 1 University, Pantheon Sorbonne, France.

Anne M. Larson is Senior Associate with the Center for International Forestry Research and is based in Nicaragua. Her research has focused on conservation and development, decentralization, indigenous rights and forest governance. She holds a PhD in wildland resource science from the University of California–Berkeley and a bachelor's degree in environmental science from Stanford University.

Emmanuel Marfo, Ghanaian, is Research Scientist (forest policy and law) at the Forestry Research Institute of Ghana. He obtained his PhD and MSc degrees from Wageningen University, The Netherlands, and his bachelor's degree in natural resources management from the University of Science and Technology, Ghana. His research interest is forest politics and governance, focusing on legal complexities, conflict, power, representation and negotiation, with a special focus on communities.

Iliana Monterroso, Guatemalan citizen, is Researcher at the Latin American Faculty of Social Sciences (FLACSO-Guatemala) and co-coordinator of the work in Latin America for Rights and Resources Initiative. She is currently finishing her PhD in environmental sciences at the Autonomous University of Barcelona, Spain. Her main areas of work are related to community forestry, socioeconomic assessment of biodiversity risks and socioenvironmental conflicts.

Pablo Pacheco, Bolivian, is Scientist at the Center for International Forestry Research. He holds a PhD in geography from the Graduate School of Geography at Clark University, in Massachusetts, USA. His main areas of work are related to agrarian change, rural development, forest policy, community forestry, livelihood strategies and landscape change.

Naya S. Paudel, Nepalese citizen, is working with ForestAction, Nepal. He received his PhD in political ecology of nature conservation from the University of Reading, UK. His main areas of work include livelihoods, environmental governance, community-based natural resources management and institutions.

Juan M. Pulhin is Professor and Scientist II in the Department of Social Forestry and Forest Governance, College of Forestry and Natural Resources, University of the Philippines Los Baños. He has more than 27 years of experience in natural resources research, education and development at the national and international level and has published more than 50 technical papers dealing with the various aspects of natural resources conservation and climate change.

Sushil Saigal works on issues related to natural resources management, with a particular focus on forestry in India. He was head of the Natural Resource Management Division of Winrock International, India, before starting his PhD in political ecology at Cambridge University in 2005.

List of Acronyms

ACICAFOC	Coordinating Association of Indigenous and Community Agroforestry in Central America
ACOFOP	Association of Forest Communities of Petén (Guatemala)
ACRA	Associazione di Cooperazione Rurale in Africa e America Latina
ARCA	Assessoria Comunitária e Ambiental (Brazil)
ASEAN	Association of Southeast Asian Nations
BOD	board of directors
BOLFOR	Bolivia Sustainable Forest Management Project (Bolivia)
CAPRi	Collective Action and Property Rights
CATIE	Centro Agronómico Tropical de Investigación y Enseñanza (Costa Rica)
CBFM	Community-Based Forest Management
CCARC	Caribbean and Central American Research Council
CEDLA	Centro de Estudios para el Desarrollo Laboral y Agrario (Bolivia)
CENRO	Community Environment and Natural Resources Office
CF	Community Forest
CFM	Community Forest Management
CFUG	Community Forest User Group (Nepal)
CGIAR	Consultative Group on International Agricultural Research
CIFOR	Center for International Forestry Research
CIRABO	Union of Indigenous People of the Bolivian Amazon
CNS	Conselho Nacional dos Seringueiros (Brazil)
COICA	Coordinator of Indigenous Organizations of the Amazon Basin
COINACAPA	Integrated Agroextractivists of Pando Farmers' Cooperative, Ltd. (Brazil Nut Producers' Cooperative, Bolivia)
CONAP	National Commission for Protected Areas (Guatemala)
COPNAG	Central Organization of Native Guarayos Peoples (Bolivia)
DENR	Department of Environment and Natural Resources (Philippines)
FAO	Food and Agriculture Organization
FAUSAC	Facultad de Agronomía/ Universidad San Carlos (Guatemala)
FECOFUN	Federation of Community Forestry Users, Nepal

FGF	Forest Governance Facility
FLASCO	Facultad Latinoamericana de Ciencias Sociales
FPIC	Free Prior and Informed Consent
FMB	Forest Management Bureau
FRCD	Forest Resources Conservation Division
FRDD	Forest Resources Development Division
FSC	Forest Stewardship Council
FSLN	Sandinista political party (Nicaragua)
FUNAI	National Foundation for Indians (Brazil)
GACF	Global Alliance of Community Forestry
GTZ	Gesellschaft für Technische Zusammenarbeit
ha	hectare
IBAMA	Brazilian Environmental Agency
ICCO	Interchurch Organisation for Development Co-operation
ICRAF	International Center for Agroforestry Research (now known as World Agroforestry Center)
IDRC	International Development Research Centre (Canada)
IFMA	Industrial Forest Management Agreement
IIED	International Institute for Environment and Development
ILO	International Labour Organization
INEC	National Institute of Statistics and Census (Nicaragua)
INRA	Instituto Nacional de Reforma Agraria, National Institute of Agrarian Reform (Bolivia)
IPCC	Intergovernmental Panel on Climate Change
IUCN	International Union for the Conservation of Nature
IUFRO	International Union of Forestry Research Organizations
KEF	Kalahan Educational Foundation (Philippines)
LAET	Laboratório Agroecológico da Transamazônica (Brazil)
Lao-PDR	Lao People's Democratic Republic
LGU	local government unit
LI	Legal Instrument (Ghana)
LIFE	Livelihoods, Income, Forest Condition and Equity
NGO	Non-Governmental Organization
NPA	New People's Army (Philippines)
NTGCF	National Tree Growers' Cooperative Federation (India)
NTPF	Non-Timber Forest Product
PAID	Panafrican Institute for Development (Ghana)
PENRO	Provincial Environment and Natural Resources Office
PROFOR	Program on Forests of the World Bank
RAAN	North Atlantic Autonomous Region (Nicaragua)
RECOFTC	Regional Community Forestry Training Centre (Thailand)
RED	regional executive director
REDD	Reducing Emissions From Deforestation and Forest Degradation
RESEX	Extractive Reserve (Brazil)
RRI	Rights and Resources Initiative
RUP	Resource Use Permit (Philippines)

SEMARNAP	Secretariat of Environment, Natural Resources and Fishing (Mexico)
SNV	Netherlands Development Organisation
SOAS	School of Oriental and African Studies
SRA	Social Responsibility Agreement (Ghana)
TCO	Tierra Comunitaria de Origen, original community land (Bolivia)
TGCS	Tree Growers' Cooperative Society programme (India)
UDAPE	Unidad de Análisis de Política Económica (Bolivia)
UDEFCO	Union Départmental des Forêts Communautaires de l'Océan (Cameroon)
UFPA	Federal University of Pará (Brazil)
URACCAN	Universidad de las Regiones Autónomas de la Costa Caribe de Nicaragua
USAID	United States Agency for International Development
VAIPO	Viceministerio de Asuntos Indígenas y Pueblos Originarios (Vice-ministry of Indigenous and Original People's Affairs) (Bolivia)
WWF	World Wildlife Fund (also known as World-Wide Fund for Nature)

Index

Franklin Pierce University
00187246